Lecture Notes in Computer Science 1014

Edited by G. Goos, J. Hartmanis and J. van Leeuwen

Advisory Board: W. Brauer   D. Gries   J. Stoer

Springer
Berlin
Heidelberg
New York
Barcelona
Budapest
Hong Kong
London
Milan
Paris
Santa Clara
Singapore
Tokyo

Angel Pasqual del Pobil
Miguel Angel Serna

Spatial Representation and Motion Planning

Springer

Series Editors

Gerhard Goos, Karlsruhe University, Germany

Juris Hartmanis, Cornell University, NY, USA

Jan van Leeuwen, Utrecht University, The Netherlands

Authors

Angel Pasqual del Pobil
Department of Computer Science, Universitat Jaume I
E-12080 Castellón, Spain

Miguel Angel Serna
C.E.I.T. & Universidad de Navarra
E-20009 San Sebastián, Spain

Cataloging-in-Publication data applied for

Die Deutsche Bibliothek - CIP-Einheitsaufnahme

DelPobil, Angel Pasqual:
Spatial representation and motion planning / Angel Pasqual
DelPobil ; Miguel Angel Serna. - Berlin ; Heidelberg ; New
York ; Barcelona ; Budapest ; Hong Kong ; London ; Milan ;
Paris ; Tokyo : Springer, 1995
  (Lecture notes in computer science ; 1014)
  ISBN 3-540-60620-3
NE: Serna, Miguel Angel:; GT

CR Subject Classification (1991): I.2.9, I.3.5, I.2.8, I.2.10, J.6, J.2

ISBN 3-540-60620-3 Springer-Verlag Berlin Heidelberg New York

© Springer-Verlag Berlin Heidelberg 1995
Printed in Germany

Typesetting: Camera-ready by author
SPIN 10512245   06/3142 – 5 4 3 2 1 0   Printed on acid-free paper

To Marián, for her unconditional love and understanding.
To Angel, María and Javier for making us so happy.

Angel P. del Pobil.

To Eduardo and Elizabeth Bayo, for their hospitality while I stayed in Santa Barbara and their warm friendship.

Miguel Angel Serna.

Preface

The work presented in this book is based on the results of the research of the authors during the last years at the *Robotic Intelligence Group* of Jaume I University and at the *Advanced Robotics and AI Laboratory* of C.E.I.T. and the University of Navarra. However, the topic of spatial representation and motion planning was first discussed by the authors while they were at the University of California at Santa Barbara in 1988.

We consider spatial representation and motion planning as a substantial part of Artificial Intelligence (AI) and Robotics. The AI community has been traditionally devoted to vision, language and different kinds of reasoning. However, if AI is to deal with real-life problems, intelligent systems will have to interact with the world in the way persons do: by moving around and manipulating things. Those are the main goals of robotics.

A good representation must be simple enough and, at the same time, contain all the necessary information to deal with the problem at hand. In the case of spatial representation for motion planning a trade-off is needed: too-simple models may be inappropriate to solve the motion planning problem, and too-accurate representations often lead to inefficient algorithms in complex real-world scenarios.

In this work we present a spatial representation that uses the simplest geometric object —the sphere— together with a hierarchical representation, to model three-dimensional objects. The model is composed of two representations: an exterior representation, made up of outer spheres, provides an upper bound limit; an interior representation, made up of inner spheres, is used as a lower bound limit. Each representation can be refined as required, hierarchically converging towards the real object.

By combining the simplicity of the sphere and the powerful hierarchical representation, the issues of obstacle-free spaces and motion planning can be addressed in a very simple manner. The algorithms make use of the fact that the sphere is the only 3D object that has only three degrees of freedom. Besides the simplicity of dealing with spheres, the hierarchy of detail allows the tuning up of the model to the required accuracy in each particular case.

The content of this book can be of interest for graduate students and researchers in the fields of AI and robotics. Those lecturers working in motion planning will find it very suitable for seminars and group discussions. Nevertheless, since most of the problems and concepts are presented in a simple and clear language, undergraduate students should not find many difficulties in understanding its content.

October 1995,   Angel Pasqual del Pobil
                                                       Miguel Angel Serna

Acknowledgments

During the development of the research presented in this book we have dealt with many people. We wish to thank all of them for their support, comments and suggestions.

Begoña Martínez Salvador deserves a special mention for her work in Chapter 5, of which she is actually co-author.

Particular gratitude is due to Professor Jose María Bastero, Vice Rector of The University of Navarra, and Professor Eduardo Bayo, at The University of California in Santa Barbara, who have encouraged us since the beginning of our work.

Professors Francisco Michavila, Carlos Conde and Lola Rodrigo, of Jaume I University, and Manuel Fuentes, Director of the C.E.I.T. Research Center, provided the necessary support to create the *Robotic Intelligence Group* and the *Advanced Robotics and AI Laboratory* at their respective institutions.

We want to thank the support of Diputación Foral de Guipúzcoa, Comisión Interministerial de Ciencia y Tecnología (CICYT, project TAP92-0391-C02-01), Fundació Caixa Castelló (projects B-41-IN, A-36-IN), Generalitat Valenciana and The British Council.

Last, the first author owes a great debt of thanks to his family for their support and understanding. Thanks to Marián for giving so much, for her patience, support and encouragement. Thanks to Angel, María and Javier for loving their father so much.

"Do not despair. Remember there is no triangle, however obtuse, but the circumference of some circle passes through its wretched vertices"

S. Beckett, Murphy

1938.

Contents

Chapter 1

Introduction

For people, manipulation and motion are the most common means of acting directly on the world. Vision provides information about our environment while language and speech serve us to interact with other human beings, but we act on our surroundings by moving ourselves or by producing the motion of objects: by grasping, carrying, pushing, twisting, etc. Even the simplest task requires a complicated combination of movements. The Artificial Intelligence community has been traditionally devoted to vision, language and different kinds of reasoning. However, a much deeper understanding of the motion problem is needed. If Artificial Intelligence (AI) is to deal with real life problems, intelligent systems will have to interact with the world in the way persons do: by moving and manipulating. If a robot is defined as a machine with the ability of moving and/or manipulating, the problem we are dealing with can be called *robot motion planning* (RMP). This book is a contribution to the study of adequate spatial representations to deal with movement in the framework of Robotics and AI applications.

1.1 Motivation and Statement of the Problem

The issue of representation is of crucial importance to Artificial Intelligence [Charniak and McDermott, 1985], [Davis, 1990]. Depending on how a representation is selected, the solution to a problem can be quite simple or very complex. In the context of spatial representations, several different models for 3D objects have been used mainly in the domain of computer graphics applications: Constructive Solid Geometry, Boundary Representation, spatial occupancy enumeration, cell decomposition, swept volumes, octrees or even prism-trees. These models are usually too graphics-oriented to be adequate as a paradigm for reasoning systems: they lack expressiveness and the elicitation of higher features of an object from these models is a very difficult task.

Generally, a trade-off between accuracy and simplicity is necessary, since a too detailed model may often be too complex, while a too simple model may be too inaccurate. A good representation in Artificial Intelligence must be one that is simple enough but, at the same time, one that contains all the necessary information to deal with the problem at hand and can be refined as required. There are two possibilities to simplify the representation of an object: decomposing the object as a combination of simpler parts, on the one hand, or computing an approximation of the shape of the object, on the other. According to Chazelle [1987], when we come to consider the situation of research regarding problems about approximation and decomposition of shapes in three dimensions, we find that relatively little is known. The mathematical theory of packing and covering [Rogers, 1964] deals with related problems; although it has a rather theoretical interest, its study is very useful as it provides us with a formal framework on which an object model may be built up.

The sphere is the simplest of all geometric objects, this is the reason why in many cases a problem can be solved when we deal only with spheres but a solution is lacking, or the problem becomes much harder, when objects with a more complex geometry are involved. Two main features of the sphere account for its simplicity: first, a sphere can be perfectly described by means of only one parameter, namely its radius; second, the motion of a sphere can be completely described by the motion of just one point, namely its center. The sphere is, by its own nature, the only free body that has three degrees of freedom in 3D; as it were, it only knows about translations.

This simplicity and elegance of the sphere —and the circle, its minor relative— have attracted a great deal of research attention both in pure and applied science. However, the applications of spheres for spatial representations in AI and Robotics are rather scarce and limited to circles on the plane or simple cases in 3D space. Some object representations based on spheres have been previously applied to collision avoidance and detection, but in most of them one of two possible simplifications is made: either the problem is formulated from the start as being limited to spheres without considering other actual objects, or a certain fixed approximation for every object is defined using a few spheres without any further possibility of improving its accuracy.

In this book, we present an investigation that tries to take advantage of the ample possibilities of spheres for spatial representations in Robotics and AI applications. By combining the simplicity of the sphere and the power of the notion of hierarchy of detail, a new practical model is proposed that is based on a double spherical representation of solid bodies. This representation is general and simple at the same time, for the sphere is the only geometric body that is used in the whole model; and that without loss of accuracy, since a representation can be as exact as desired. For example, a real robot, like the one shown in Fig. 1.1, would be represented as the set of spheres in Fig. 1.2. Moreover, the model is twofold: an exterior representation —composed of outer spheres— provides us with an excess approximation, while an interior representation —composed of inner spheres— is used as a defect approximation. These inner spheres are contained within the object, and they will prove very useful to speed up the process of detecting collisions. We will present first the spatial representation model, and will apply it later to the motion problem in AI, in particular to collision detection and motion planning for robot manipulators in three dimensions.

Motion planning is a fundamental issue for Artificial Intelligence applications in Robotics [Brady et al., 1984]. Its main aim would be to endow an autonomous robot with some basic capabilities to generate its own motion. A first condition for such movement would be to avoid collisions with possible obstacles that may be found in its path. Even though most superior animals possess this ability, which in principle does not require very much intelligence, the solution to the problem is by no means easy. The reason for this apparent contradiction is found in the fact that the coordination of our own movement is done mainly in an unconscious way —by hardware, so to speak— depending a great deal on our perception systems without a logical deductive process coming into play [Hollerbach, 1982].

The collision-free planning of motion can be generally stated as the problem of developing algorithms to automatically compute a continuous safe path for a given set of objects (possibly linked), in such a way that they move from an initial placement to a final placement avoiding collisions with obstacles.

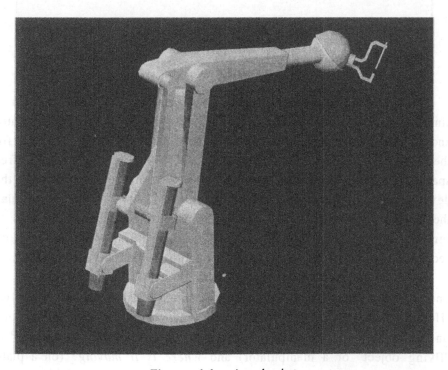

Figure 1.1 A real robot

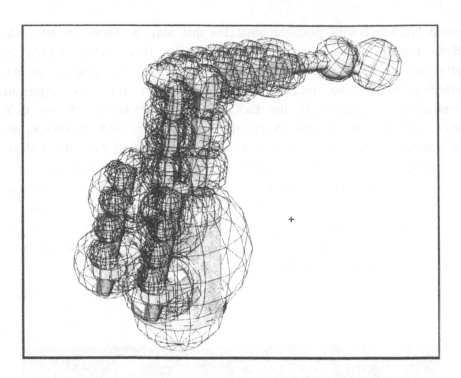

Figure 1.2   Spherical representation for the robot shown in Fig. 1.1

For a robot manipulator, this problem can be formulated in a simpler way by saying that, given a description of the manipulator and a known environment, a collision-free path has to be planned for the robot arm between two positions, i.e., it must move inside free space. The solution to this problem is a preliminary step towards the development of higher level robot planning that will allow us to plan the actions of robots at *task level*; that means that we will be able to ignore the implementation details and the particular motion sequence that the robot must follow to carry out a certain task.

It can be stated without exaggeration that motion planning is one of the most complex problems in the domain of Robotics [Hopcroft and Krafft, 1987]. Moreover, its complexity depends mainly on the answer to two essential questions, namely: *what is moving?* (a "flying object" or a manipulator) and *where is it moving?* (on a plane or in 3D space). Changing from two to three dimensions makes the problem inherently more difficult (see Fig. 1.3), and it becomes even much harder for robot manipulators as the number of degrees of freedom increases (see [Brooks, 1983] for a longer discussion of this point).

Collision detection or the problem of how to recognize a safe path is a previous question that has to be solved prior to the motion planning problem. Obviously, we must be able to assure that a given path is safe before searching for a sequence of collision-free motions going from the start to the goal.

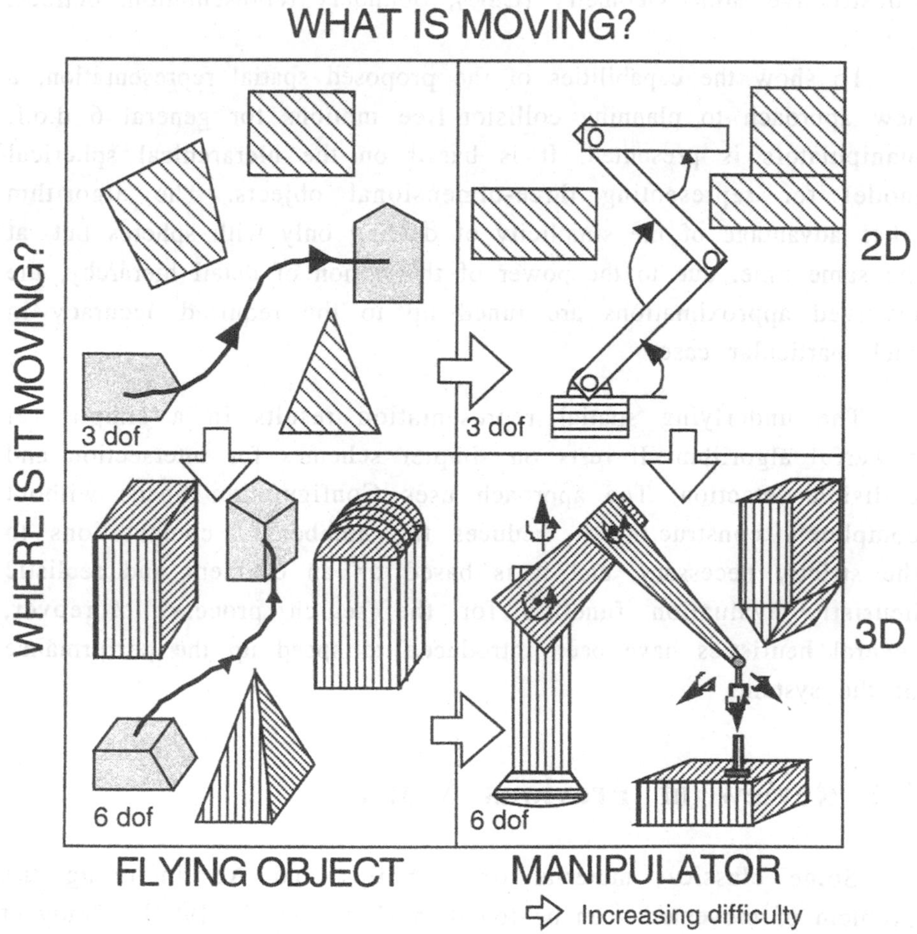

Figure 1.3 Difficulty of the motion planning problem

Two main issues can be distinguished when solving the motion planning problem. The first one is how to construct the space of Free Placements (FP) for the robot in its workspace. The second issue is how to search for a solution path inside the FP space. The first question is directly dependent on the selected collision detection scheme, since many candidate motions will have to be checked to determine whether they are valid. In addition, this is highly influenced by the particular object model that is used by the off-line planner to represent the involved objects (obstacles and robot links): Constructive Solid Geometry (CSG), boundary representation, octrees, etc.

To show the capabilities of the proposed spatial representation, a new approach to planning collision-free motions for general 6 d.o.f. manipulators is presented. It is based on the hierarchical spherical model for representing three-dimensional objects. The algorithm takes advantage of the simplicity of dealing only with spheres but, at the same time, due to the power of the notion of detail hierarchy, the involved approximations are tuned up to the required accuracy in each particular case.

The underlying spatial representation results in a simple yet powerful algorithm. It rests on simpler schemes for intersection and collision detection. The approach uses Configuration Space without completely constructing it, reduces the number of computations to the strictly necessary, and it is based on an efficient and realistic heuristic evaluation function for the search process. Moreover, several heuristics have been introduced to speed up the performance of the system.

1.2 Review of Previous Work

Some classical material on Robotics and AI, including the problem of movement, can be found in [Brady et al., 1982], [Brady et al., 1984], [Brady, 1985], [Brady 1989], [Khatib et al. 1989], [Cox and Wilfong, 1990], the last one focusing mainly on mobile robots and MP on the plane. For a good state of the art on Robotics [Albus, 1984] and [Korein and Ish-Shalom, 1987] can be mentioned, and [Hopcroft and Krafft, 1987] is an excellent summary of future directions in Robotics research from the point of view of Computer Science.

1.2.1 Spatial Representations

Good reviews of the various representations for 3D objects that have been used mainly in the domain of computer graphics applications can be found in [Reddy and Rubin, 1978], [Baer, Eastman and Henrion, 1979] and [Requicha, 1980]. Octrees are introduced in [Samet, 1984] and prism-trees, another interesting representation, is described in [Faugeras and Ponce, 1983]. A discussion of swept volume techniques can be found in [Wang and Wang, 1986]. An interesting representation paradigm based on generalized cones with good results for elongated objects is presented in [Agin and Binford, 1976]. The hierarchical nature of CSG has made possible some important results: Lee and Fu [1987] extract manufacturing features from it, Cameron [1989] uses it for intersection detection and Faverjon [1989] for collision avoidance.

Particular models have been developed for specific AI applications: namely [Brady, 1984] and [Marr and Vaina, 1980] must be mentioned for their models for representing the shape of objects, as well as [Faugeras et al., 1984] in the context of artificial vision. A general survey of representations of space for commonsense reasoning can be found in [Davis, 1990]. In a previous paper Davis [1988] deals with the problem of how to describe and predict the behavior of solid objects, reaching very interesting conclusions.

As it has already been stated, the simplicity and elegance of the sphere have motivated its use in many different fields. To mention but a few, [Connolly, 1987] deals with spheres in the domain of molecular biology; Lee and Preparata [1984] surveyed computational geometry where problems about spheres are very frequent; Avis, Bhattacharya and Imai [1988] studied the computation of the volume of a set of intersecting spheres, Hopcroft, Schwartz and Sharir [1983b] proposed an algorithm to detect intersections in a set of spheres and Sharir [1985] for a set of disks; in the mathematical theory of packing and covering the particular case of sets composed of spheres has been studied in depth [Bambah, 1954], [Barnes, 1956], [Few, 1956], [Coxeter, Few and Rogers, 1959], [Rogers, 1964]; finally, in the framework of geometric location theory [Francis and White, 1974] related problems are dealt with, as the so-called p-center problem [Megiddo and Supowit, 1984].

1.2.2 Collision Detection and Motion Planning

Many previous approaches to collision-free motion planning or other related problems in Robotics take advantage, in one way or another, of the easiness of working with spheres. Most of them, however, are based on a fixed, rather coarse approximation that cannot be improved in any way. Bajaj and Kim [1988] have studied a unique moving sphere, while Chen [1990] used a spherical model for the perceptual space. Early works in path-planning made a rather simple use of spheres by merely enclosing every object inside one sphere [Pieper, 1968], [Widdoes, 1974].

More recent approaches utilize a certain set of spheres or circles to model the robot or the obstacles [Abramowski, 1988], [Singh, 1988], [Thakur, 1986], [de Pennington, Bloor and Balila, 1983], [Esterling and Van Rosendale, 1983], [Tornero, Hamlin and Kelley, 1991]. In the case of mobile robots, many approaches have been proposed that approximate the robot and the obstacles by enclosing them in a circle [Kambhampati and Davis, 1986], [Moravec, 1981], [Thorpe, 1984], [Ichikawa and Ozaki, 1985]. There also exist other publications dealing with special treatments of geometric planning problems in which the moving objects are circles [O'Dúnlaing and Yap, 1985], [Schwartz and Sharir, 1983b], [Yap, 1984], [Spirakis and Yap, 1984].

Perhaps the most closely related antecedent to our spherical model is the work of Badler, O'Rourke and Toltzis [1979] who associated the concept of level with a spherical approximation for collision detection, but this model extends the representation to only two levels. More recently, Bonner and Kelley [1990] developed a system that uses a sequence of rectangular sectors of spheres to efficiently find a path for a single object moving in straight lines in 3D space.

Hierarchy has proved to be a very useful tool in solving the problem that concerns us. Zhu and Latombe [1991] have successfully applied the notion of hierarchy to efficiently solve the problem of path planning in 2D. Faverjon and Tournassoud [1988] have also applied this notion to motion planning; in this approach collision detection is based on an octree representation, as it is the case for other approaches that have been lately presented. Octrees, however, present important drawbacks when dealing with motion [Dupont, 1988], [Hayward, 1986], as the involved transformations (rotations and translations) are computationally very expensive, since the complete representation tree has to be computed anew for each placement of the robot. Cameron [1989, 1990] combines the hierarchical structure of a CSG scheme with a four-dimensional intersection testing approach to collision detection obtaining very interesting results, even though the method involves certain mathematical complexity.

Boyse [1979] proposed a method to detect collisions that was based on an implicit representation of swept volumes in order to avoid the complexity of explicitly computing swept volumes. Taking Boyse's work as a starting point, an approach that characterizes the different types of collisions between vertices, edges and faces has been widely used (the most remarkable references are [Lozano-Pérez, 1987], [Canny, 87], [Donald, 1987] among many others). It is based on a boundary representation scheme and computes collisions by checking all possible intersections between the edges, vertices and faces of all objects. Although this approach yields good results for simplified models of reality, its performance may be hampered when applied to actual situations in which the design of objects presents a complex geometry with thousands of vertices. A rather different approach makes use of quaternions as a tool for representing the problem [Canny, 1986].

Many other references can be mentioned that are relevant to motion planning, some of them are included here. For a more exhaustive review of previous work see [Latombe, 1991], [Hwang and Ahuja, 1992]. Most approaches to collision detection reduce the problem to an instance of an intersection detection problem; Shamos and Hoey [1976], Bentley and Ottmann [1979] and Chazelle and Edelsbrunner [1988] use plane-sweep techniques for computing intersections on the plane. The question of detecting an intersection between two polyhedra has been studied by Ahuja et al. [1985] and Dobkin and Kirkpatric [1985], while Edelsbrunner [1982] and Hopcroft, Schwartz and Sharir [1983b] deal with intersections among several objects, two sets of rectangular parallelepipeds in the first case, and a set of spheres in the second. For objects in motion, the multiple intersection test is usually applied in the domain of computer graphics: Uchiki, Ohashi and Tokoro [1983] describe a typical system based on spatial occupancy enumeration; Hayward [1986] uses octrees for representing objects; Moore and Wilhelms [1988] present two approaches, the first one using multiple intersection test and boundary representations and the second applying implicit sweep volume for a cell decomposition of each surface into triangles.

   In the domain of motion planning, much work has been done for polygons moving on the plane: [Lozano-Pérez and Wesley, 1979], [Lozano-Pérez, 1983], [Brooks and Lozano-Pérez, 1985] and [Brooks, 1983a] are seminal papers on this subject, in the second one the notion of Configuration Space is first applied to motion planning. Buckley [1989a] presents a rather different approach by using constrained optimization techniques. For polyhedra moving in 3D space Wong and Fu [1985] consider a moving object with only three degrees of freedom, while Donald [1987] studies the motion of a polyhedron with six degrees of freedom. An algorithm called *roadmap* has been presented in [Canny, 1988], it is very interesting from the point of view of complexity theory and it improves on a previous general algorithm by Schwartz and Sharir [1983].

   In the case of manipulators, simplifications must be introduced for practical applications due to the inherent complexity of the problem: [Kantabutra and Kosaraju, 1986], [Chien, Zhang and Zhang, 1984] and [Chen and Vidyasagar, 1987] deal with robots represented as a set of linked rods moving on the plane. In [Ozaki, 1986] all intervening objects are constrained to be rectangular parallelepipeds. Gouzènes [1984] first points out the need for global planning and for an explicit construction of free space, the implemented model is a planar robot with 2 or 3 d.o.f. moving in an environment of rectangles.

An interesting approach to manipulators in 3D is that of Brooks [1983b], who reduces the number of degrees of freedom of a PUMA arm to four and restricts the possible motions of the robot as well as the set of possible obstacles. Lozano-Pérez [1987] has proposed a technique called *slice projection* to build the configuration space for a manipulator: the range of possible values for each joint is discretized and the links are accordingly enlarged to assure that no collision occurs for the coordinate values that are not computed, the collisions are detected by checking all possible intersections between vertices and edges (or vertices, edges and faces in three dimensions). This method works well for three or four d.o.f., but if the number of vertices in the model is large or there are more than four degrees of freedom and the robot links are rather long, lengthy computations may be necessary or even valid interesting paths may be lost due to the involved approximations.

A quite different approach is that of the so-called *potential field* method [Khatib, 1986]. It must be regarded as a technique for collision avoidance rather than for motion planning: obstacles contribute in a negative way to the field while the goal has a positive contribution. Khosla and Volpe [1988] have suggested an alternative potential function. Canny and Lin [1990] have obtained good practical results by combining a simplified version of the roadmap algorithm [Canny, 1988] with techniques based on a potential field method.

Another research area within the field of motion planning is *geometric* or *algorithmic motion planning*. Its objective is to analyze the complexity of exact theoretical algorithms for motion planning problems. These approaches are extremely interesting from the point of view of algorithmics and computational geometry, but they cannot usually be directly implemented. A review of these algorithms can be found in [Yap, 1987], [Schwartz and Sharir, 1988] and [Sharir, 1987]. A collection of papers on these topics has been published as [Schwartz, Hopcroft and Sharir, 1987]. Additional references are:

[Reif, 1979], [Hopcroft, Joseph and Whitesides, 1984 and 1985], [Spirakis and Yap, 1984], [Schwartz and Sharir, 1983a], [Schwartz and Sharir, 1984], [Leven and Sharir, 1987a], [O'Dúnlaing, Sharir and Yap, 1983, 1986 and 1987], [O'Dúnlaing and Yap, 1985], [Kedem and Sharir, 1986], [Avnaim, Boissonat and Faverjon, 1988], [Tannenbaum and Yomdin, 1987], [Canny, 1988].

Several references concerning variants of the classical motion planning problem, as it has been defined here, are just mentioned now to complete the present survey.

Coordinated motion. Two or more independent robots move simultaneously and their motions must be coordinated to avoid collisions among them or to cooperate in a certain task. Some references that deal with these problems are: [Buckley, 1989b], [Erdmann and Lozano-Pérez, 1987], [Faverjon and Tournassoud, 1986], [Hopcroft, Schwartz and Sharir, 1984], [Lee and Lee, 1987], [Parsons and Canny, 1990], [Schwartz and Sharir, 1983b], [Yap, 1984], [Spirakis, 1984].

Moving obstacles. Obstacles are allowed to move, generally following known paths with known velocities. As speed must now play an important role, most approaches take it as an additional coordinate: [Fujimura and Samet, 1989], [Kant and Zucker, 1990], [Maciejewski and Klein, 1985], [O'Dúnlaing, 1987], [Shih, Lee and Gruver, 1990].

Non-holonomic constraints. In this case the robot movements are restricted by non-holonomic constraints: some of their equations are not algebraic, that is, they are expressed in terms of the derivatives of some coordinates. This situation arises frequently for mobile robots. References for this kind of problems are: [O'Dúnlaing, 1987], [Jacobs and Canny, 1990], [Tournassoud and Jehl, 1988].

Motion planning under uncertainty. If the environment is only partially known, or errors are introduced resulting in an uncertainty about the exact position of objects, then special techniques must be used for motion planning: [Erdmann, 1984], [Donald, 1988], [Brost, 1986], [Erdmann and Mason, 1986], [Latombe, Lazanas and Shekhar, 1991].

Fine motion and compliant motion. When objects are in contact
[Hopcroft and Wilfong, 1986] or very close to a contact position, very
precise movements are required which are usually beyond the
controller's capabilities [Mason, 1984], [Juan and Paul, 1986].
Particular approaches are then necessary [Lozano-Pérez, Mason and
Taylor, 1984], one of the most important being compliant motion
[Mason, 1982].

Chapter 2

The Spatial Representation

The purpose of this chapter is to study the possibilities of the sphere as a paradigm for spatial representation in AI and Robotics. Although the proposed object model will be used later in the context of collision detection and motion planning, it is potentially appropriate for other problems in AI, as it will be discussed in the section about future research.

The possibility of using a twofold approximation based on spheres for motion planning was first suggested by Hopcroft and Krafft [1987] in a speculative paper on the future challenges of Robotics for Computer Science. Our model was conceived as we realized that many problems could usually be solved for the particular case of spheres (see references in section 1.2.1) but they became much harder in a general case. The main drawback of previous related approaches was that spherical approximations tend to be coarse. Our model combines the simplicity of dealing solely with spheres with the power of a hierarchy of detail that permits us to quantify and control the accuracy of the representation. For this purpose several heuristic techniques have been implemented (see also [del Pobil and Serna, 1992]). Its main features are as follows:

(i) The only elements that take part in the representation are spheres; therefore, any problem is always reduced to dealing with sets of spheres. The particular nature of the object underlying the representation can be ignored.

(ii) The model is always composed of two sets of spheres: the spheres in the first set —the exterior representation— cover the outer surface of the object, while the spheres in the second set —the interior representation— are contained in the object.

(iii) A hierarchy of sets is defined for both representations so that the approximations for the object —for its outer boundary or its interior— can be made better and better by using more and more spheres. For each representation a set of parameters is introduced which provide us with a global and local measure of the degree of accuracy of the model. Moreover, this hierarchy converges to a zero-error representation.

Therefore, the first step to build up our representation is to *spherize* all the objects that are relevant to our particular problem. That amounts to transforming every involved object into two sets of spheres. The first set is called the exterior representation and has the property of enclosing the boundaries of all objects. The second set of spheres forms the interior representation. Each interior sphere must be contained in one of the objects being represented. To summarize, the outer spheres enclose the boundary of an object and the inner spheres are enclosed within the object.

In addition, both of these representations must have certain desirable properties. First, they have to be *improvable*, that is, a better representation can be defined to substitute the given one. Second, a representation must be balanced: its error zone cannot be concentrated around a location; instead, it should be homogeneously scattered. Finally, a representation should be optimal to a certain degree; that is, a better representation cannot be found using the same number of spheres.

Efficiency will not be considered a crucial factor at this stage. All representations are computed only once and stored for later use. As the *spheration* process may be included as part of the complete design process, the computer time needed to automatically generate all representations of a set of objects will always be negligible when compared with the time —and effort— spent on the rest of the process by a person using a geometric modeler, however skilled he or she may be.

Before proceeding with the details of the *spheration* process, it has to be pointed out that this process has an important heuristic component in its conception. This must be so due to the inherent complexity of the problem. Even for the simpler planar case it has been shown to be NP-hard [Megiddo and Supowit, 1984]. One could conceive other valid solutions to our representation problem for a fixed number of spheres. These solutions would be valid in the sense that they should be properly defined — as interior or exterior representations— and should have the formerly mentioned good features. The proposed solution is valid and is the only one known so far by the authors. Related work can be found in [del Pobil, Serna and Llovet, 1992] and [del Pobil and Serna, 1994a]

2.1 Exterior Representation

Given a set of m geometric objects and an integer n we wish to define a certain set of n spheres in such a way that there is not any point P belonging to the boundary of an object that is not contained inside at least one of the spheres in the representation. That is to say:

$$\forall\, P \in BO_i \ \exists\, S_j \in S \mid P \in S_j$$

Where:
$O = \{O_i,\ i=1,...,\ m\ \}$ is the set of all objects.
$S = \{S_j,\ j=1,...,\ n\ \}$ is the set of all spheres in the representation.
BO_i is the boundary of object O_i.

2.1.1 Planar Case

Let us first consider the particularization of the same problem to 2D space. Our objects will now be convex polygons. The demand for convexity limits to a certain extent the application field of the method. There exist well-known techniques that have been developed in the domain of computational geometry to solve the problem of finding a partition of any polygon in a set of convex polygons (for example, [Chazelle, 1987], [Lee and Preparata, 1984] or [Preparata and Shamos 1985] can be consulted), but this is not possible for polygons that are not locally non-convex. In addition, the problem becomes more complex if the optimal partition is desired, that is, if it should contain the least possible number of convex polygons. In chapter 5 an extension of our method to non-convex generalized polygons is discussed.

a) One circle

The first outer representation for a polygon is obtained by enclosing it within a single circle. For this purpose it suffices to find a disk that covers all vertices of the polygon. To be an optimal representation this disk must be the smallest one that encloses all vertices. This definition of the problem corresponds to the 1-center problem in location theory [Francis and White, 1974] and can be solved in $O(n\log n)$ time by means of an algorithm proposed by Shamos and Hoey [1975] which utilizes the farthest neighbor Voronoi diagram. This complexity has subsequently been improved to optimal $O(n)$ by Megiddo [1983] in a paper on linear programming in $\Re^3$. This last method, however, is not practical when applied to problems of modest size [Chazelle, 1987]. As it is based on linear-time median computation, its performance may be worsened if n is not large enough. Consequently, Megiddo's method has been discarded, since, in our case, the number of vertices is usually expected to be less than 20.

Shamos and Hoey's algorithm has not been employed either in order to avoid the difficulty of constructing the Voronoi diagram, while keeping in mind that efficiency is not a crucial point for our problem. A simpler method has been used instead. It is based on the fact that the smallest enclosing circle passes through at least two vertices of the polygon. The method looks for the vertices that define the smallest enclosing circle by bounding the possible solutions to accelerate the search process.

First, the farthest pair of the set of vertices is computed. This couple defines the *diameter d* of the polygon and, obviously, if the solution circle only passes through two vertices its diameter will coincide with the diameter of the polygon and will be limited by the two points A and B that make up this pair. Let $R=d/2$ be the radius of this circle. If in fact, all the remaining vertices are contained in this disk, the problem would be solved. If this were not so, it would mean that the smallest enclosing circle would cut through at least three vertices of the polygon, and its radius would be bounded by $[\sqrt{3}R, R]$. Indeed, the lower bound is given by the diameter of the polygon, which makes the diameter of the desired circle be at least $2R$, because if it were less it could not contain the farthest pair. The upper bound corresponds to the radius of a circle C_{max} with the center at the midpoint between A and B, and which passes through points E and F, cutting points of two circumferences with the centers in A and B and with radius d. This circle covers all of the points but it is not optimal (Fig. 2.1). It is evident that a vertex of the polygon outside of C_{max} cannot exist since its distance from A or B would be greater than d, thus contradicting the definition of the diameter of a polygon.

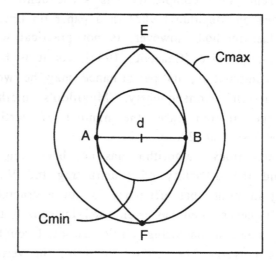

Figure 2.1   Bounds for the radius of the smallest enclosing circle

Next, a search process for a solution among all the possible trios of vertices of the polygon is initiated. Each one of them defines a circle which is a candidate for the optimal solution. First, it is verified that its radius r fulfills that

$$R < r < R_{max},$$

where initially $R_{max} = \sqrt{3}R$. If it proves otherwise, the solution is rejected. If it fulfills this condition, it is verified that all the vertices are covered. If this is so, $Rmax$ takes the value of the radius of the last accepted circle which will be the solution circle with the least radius found up to now. By repeating this process over and over again, the minimum enclosing circle will be obtained.

Even though linear algorithms for computing the diameter of a polygon exist [Bhattacharya and Toussaint, 1988], we have preferred to use —for the sake of simplicity— the direct algorithm which is $O(n^2)$. In any case, the complexity of the algorithm is dominated more by the search for the minimum circle rather than the computation of the diameter, as we shall discuss next.

In principle, the complexity in the worst case for the global algorithm would be $O(n^4)$, where n is the number of vertices. This is worse than that obtained by the above-mentioned algorithms. This complexity arises from a factor n^3 resulting from checking all trios of vertices, which, furthermore, must be multiplied by a term in n, derived from the verification that all of the points are covered for a given circle. Nevertheless, the average computation times are acceptable due to the fact that on the one hand, values of n above 20 are not expected, and also for these values we must keep in mind the exact expression for the number of trios of vertices, that is, including the constant factors and the additional terms apart from the dominant one:

$$(n^3 - 3n^2 + 2n)/6.$$

An expression which, for the worst foreseen case with $n = 20$, gives a number of operations that equals 1.140; this is perfectly acceptable, and even more so if we keep in mind that due to the fact that the bounds of the solution become more and more restrictive, the additional factor n derived to verify that all the points are covered would become more reduced.

b) More than one circle

The problem we are concerned with is classified among covering problems. In the case of using a set of circles as the generating set of the covering, the earliest result was obtained, as usual, by a mathematician [Kershner, 1939]. Later work has been presented within the framework of geometric location theory, where problems of a practical interest such as warehouse location [Shannon and Ignizio, 1970] have motivated a great deal of research. In this context, the so-called Euclidean p-center problem is a standard minimax problem, with the smallest enclosing circle problem being just its particularization for p=1. This problem poses the following question: *given n demand points in the plane, find p supply points (anywhere in the plane) so as to minimize the maximum distance from a demand point to its respective nearest supply point.* The Euclidean p-center problem has been shown to be NP-complete [Megiddo and Supowit, 1984] (see also [Kariv and Hakimi, 1979] for a more restrictive version of the problem in graph theory). In fact, even an approximate solution remains NP-hard. Accordingly, the related disk cover problem is also NP-hard: *given a set of points and disks in the plane, is it possible to arrange the disks so as to cover all the points?*

Our aim is now to find an outer representation for a polygon by means of a set of *n* disks. The radius and location of these disks must be determined in such a way that the resulting representation covers the boundary of the polygon, i.e., its edges. In addition, it must have the desirable characteristics that were mentioned above, namely, it can be improved, it is balanced, and it is quasi-optimal. A non-heuristic solution is out of the question after the precedent considerations on the complexity of related simpler problems and taking into account the fact that the number of points we are covering is not finite.

To achieve our end we will partition the boundary of the polygon to make *n* disjoint portions out of it, and we will cover each of these portions with just one circle. A portion of the boundary is defined by a set of points that will be called a *list of points* or simply *a list*. Each of the points forming a list must be a vertex of the

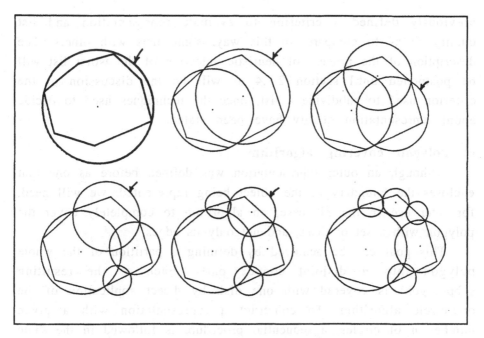

Figure 2.2   Sequence of representations showing the worst lists

polygon with the exception of the two ends of the list which may also be any point on one of the edges. To cover one of these lists with a single disk, all that has to be done is just apply the previous algorithm for the case of one circle.

The heuristic approach we present follows a sequential track to compute the exterior representation for a polygon using a given number n of disks. Such representation is obtained by modifying the representation with $n-1$ disks. The *worst* list in this previous representation is selected and subdivided into two new sublists. The new nth representation is thus formed by substituting the circle enclosing the worst old list with the two circles enclosing the new sublists. In this way we construct a whole set of representations starting with only one disk and computing a new representation from the last already obtained (Fig. 2.2). The selection of this *worst* list will be done in such a way that the new representation that comes as a result with n disks will be better than the previous one with $n-1$. Naturally, deciding which is the worst list implies having previously defined a criterion to evaluate representation and list quality so as to compare, in this way, some lists with others. The description of the process of heuristic selection of the worst list will be postponed until section 2.1.4 as will be the discussion of the criterion used to subdivide a list, once the techniques used to decide about representation quality have been stated.

c) Polygon covering algorithm

Although an outer representation was defined before as one that encloses the boundary of the object being represented, we will need, for later use in the 3D case, an algorithm to completely cover the polygon with a set of disks, and not only its edges.

This goal can be achieved by defining a partition of the whole polygon into n disjoint convex parts. Each of the resulting subpolygons is covered with one disk by direct application of the one-circle algorithm. To construct a representation with a given number n of circles, a sequential procedure is followed in the same way as the previous case: the case with n circles is obtained by improving the previous case with n-1. We begin with the first covering with a single disk, we subdivide the polygon into two parts thus obtaining two subpolygons that are covered, in turn, with two disks to form a second covering. Next, we select the *worst* of those subpolygons and we subdivide it into two more subpolygons, and so on. In figure 2.3 we can see this process for a certain polygon (note that some of its corners are rounded giving rise to more than one vertex). In each example, the circle that covers the worst polygon has been indicated.

It should be noted that the treatment of this case is very similar to the previous case for the exterior representation; more so than what it may seem at first sight. In fact, if in the preceding explanation we talked about sublists instead of subpolygons, the description of the method would be exactly the same since the algorithm for the minimum enclosing circle is as valid if the covered points are the vertices of the polygon as it is when they represent a portion of the edges of a polygon. Therefore, both methods will differ only in the criterion for selecting the worst list —or subpolygon in this case— and in the technique used to subdivide it. This fact will have important practical consequences when implementing the spherizer since the data structures and the modules will be common to both cases and the only modules that will not be shared will be the ones in charge of these two operations.

The discussion of the methods used to select the worst subpolygon and then split it in two will also be left for section 2.1.4, as it is logical that these methods will be justified by the considerations that refer to the quality of the resulting representation.

2.1.2   Extension to Three Dimensions

When we come to consider the situation of research regarding problems about approximation and decomposition of shapes in three dimensions —and covering in particular— we find that relatively little is known [Chazelle, 1987], [Rogers, 1964]. An $O(n)$ time algorithm has been obtained to compute the smallest enclosing

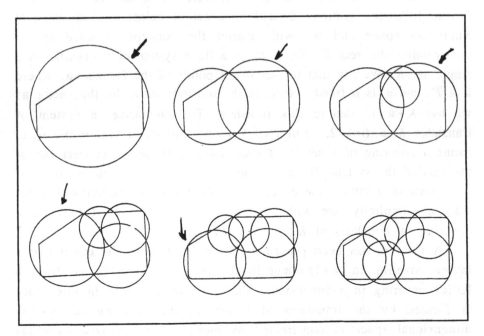

Figure 2.3   Sequence of representations showing the worst circles

hypersphere [Megiddo, 1983] as an extension of Megiddo's procedure for the case of a circle. As in the planar case, this algorithm does not yield an adequate behavior if the problem is of modest size [Chazelle, 1987], and therefore it has not been used in our approach either.

The question of covering space by means of spheres has only been dealt with in the framework of the mathematical theory of packing and covering [Bambah, 1954], [Coxeter, Few and Rogers, 1959], [Few, 1956], [Rogers, 1964]. This discipline, however, is mainly concerned with the problem of covering the complete space or a cube and not with the utilization of spheres as a medium to approximately represent any geometrical object.

a) Covering of solid objects by generalized cylinders

In the terminology of abstract set theory, the system of sets S_1, S_2, ... is said to cover the set O, if

$$\cup S_i \supset O$$

i.e., if each element of O belongs to at least one of the sets $S_1, S_2, ...$. In the following sections O will be a set of points in 3-dimensional Euclidean space and we will restrict the concept of covering by constraining the sets $S_1, S_2, ...$ to be a finite system of *translates* of a single set K. We say that the set of all points of the form $\mathbf{k}+\mathbf{a}$, where $\mathbf{k} \in K$, and $\mathbf{a}$ is a fixed vector, is the translate of K by the vector $\mathbf{a}$; we use $K+\mathbf{a}$ to denote this translate. To summarize, a system of translates $K+\mathbf{a}_i$ (i=1, 2, ...) of a set K by a sequence of vectors $\mathbf{a}_1, \mathbf{a}_2, ...$ forms a covering of a set O, if each point of O lies in at least one of the sets of the system. If the vectors $\mathbf{a}_1, \mathbf{a}_2, ...$ are an enumeration of the points of a lattice, the covering is called a lattice covering. For the sake of simplicity, we will just say *a covering with K* to mean a covering by translates of K.

A theorem has been proven by Rogers [1964] to the effect that a lattice covering in $(n+1)$-dimensional space can be obtained from a lattice covering in n-dimensional space. If the covering in this space is formed by the translates of a set C, the covering in $(n+1)$-dimensional space is constructed by means of generalized cylinders with C as the base. The present approach takes this theorem as its starting point, but it does not restrain coverings to be lattice coverings; moreover, cylinders are later substituted by spheres to render a final covering with a generating set K composed only of spheres.

The concept of a generalized cylinder is obtained from the concept of the Cartesian product of two sets. In this way, if F is a set in m-dimensional space and G is a set in n-dimensional space, the Cartesian product of F and G is defined, in the space of $n + m$ dimension, as the set of all the points that can be described as

$$(x_1, x_2, ..., x_m, y_1, y_2, ..., y_n),$$

where $(x_1, x_2, ..., x_m) \in F$ and $(y_1, y_2, ..., y_n) \in G,$

and is denoted by $F \times G$. We will call this Cartesian product a generalized cylinder since it constitutes a natural generalization of the process used to construct a cylinder.

Let us assume that the set of all bodies to be represented can be decomposed into a new set made up of simpler solid objects: $\{ O_i, i=1,...,m \}$ in such a way that each of these objects can be generated by translational sweeping of a planar convex set. That is, an object O_i must be the Cartesian product $B_i \times H_i$, where B_i is a convex polygon and H_i is a line segment perpendicular to the plane of B_i and having an endpoint on this plane. Hereafter, unless otherwise stated, we will use the term *object* to refer to this particular kind of objects.

According to our definition, an outer representation of a generic solid object O must cover its boundary surface. In the case of translational sweeeping, this *envelope* surface consists of two categories of surfaces: the surface of the 2D generator set B at the initial and final positions, and new surfaces generated during the sweeping of B. The two former surfaces will be designated as the *bases* of the object, and the latter surfaces as its *sides*. A first idea of the proposed solution scheme can be taken from the following considerations: the sides of an object can be regarded as the result of sweeping the edges of the generator polygon; therefore, if we have a set of circles that cover all the edges in the plane, and we sweep these circles, this process will produce a covering for the sides of the 3D solid object by means of cylinders with those circles as bases.

This initial result can be generalized to establish that, given a convex polygon B and a covering for its boundary with a set C, the solid generated by translational sweeping of this polygon has a straightforward covering for the side surface of its envelope surface which is composed of generalized cylinders having C as base. In a similar way, a covering for the whole swept volume can be found from a covering for the complete generator B.

b) Covering cylinders with spheres

So far we have derived a technique to compute a covering with cylinders for the sides of an object, but our aim is a final outer representation of the object using only spheres. Moreover, a further question is still unsolved, since the bases of the object —top and bottom— are not yet covered. Let us enter now upon the former issue to see how cylinders can be transformed into spheres, leaving the latter for the next section.

A set of cylinders has been defined to cover the side surfaces of a solid object. The solid has a polygon B and a line segment H as its generators. All the cylinders share the same generating segment H but each one has its own generating circle C_i. To cover just one of these cylinders with a given number n of spheres we can proceed in the following way (Fig. 2.4). Let r_i be the radius of its circle base C_i, h its height (h is the length of the segment H), A_i a point located at a distance $d=h/2n$ from the center of C_i along the segment H and let $\mathbf{a}$ be a vector with its origin at the center of C_i and its endpoint at A_i. Then a sphere S_i can be defined having its center at A_i and its radius R_i satisfying $R_i^2=r_i^2+d^2$. It can be readily demonstrated that the system of translates $S_i+\mathbf{a}_j$ (j=1, 2, ..., n) forms a lattice covering of the cylinder with S_i (relaxing notation, we have use S_i to denote the unitary set containing S_i), where the sequence of vectors $\{\mathbf{a}_j\}$ is defined from vector $\mathbf{a}$ as $\mathbf{a}_j=(2j-1)\mathbf{a}$, (j=1, 2, ..., n).

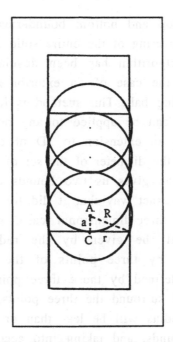

Figure 2.4   Covering for a cylinder with spheres

As an immediate corollary of the last result, the following consequence can be established. The side boundary of a translational sweeping object is covered by the system $K+a_j$ ($j=1, 2, ..., n$), where K is the set entirely composed of spheres $\{S_i, i=1,...,m\}$. The vector a_j and the sphere S_i are defined as above, each S_i being associated with a cylinder and a circle C_i of the set $\{C_i, i=1,...,m\}$ that forms a 2D covering of the generator polygon B. It has to be noted that a and the sequence $\{a_j\}$ is the same for all cylinders, although the spheres S_i have different radii. The total number of spheres in this covering is obviously $m \times n$.

A remark has to be made with respect to the case when only one sphere is used to cover the lateral surface of the object ($m=n=1$). In this case, it is a simple task to show that such a sphere encloses the whole solid. By definition, it covers the sides of the object; in fact, it encloses the cylinder defined by C_1 and this cylinder, in turn, encloses those side surfaces. But, in addition, as C_1 encloses the generator polygon of the object, the cylinder totally encloses the object, including its top and bottom boundaries, and, therefore, the sphere is by itself a covering of the entire solid object.

Furthermore, an algorithm has been developed that generalizes the previous one for the case of an exterior representation with a unique smallest enclosing ball. This method is independent of that of Megiddo's [1983] and can be applied to any polyhedron whatsoever. It can be regarded as an extension to 3D of our technique for the case of a circle. First, the diameter of the set of points defined by the vertices of a polyhedron gives us two bounds for the radius of the solution sphere in the exact way that it did for the planar algorithm. Nevertheless, for the three-dimensional case, we can find another lower bound that will be given by the radius of the greatest circumference defined by three points of the set. If we cut the sphere by the plane defined by those three points we will obtain a circumference that will surround the three points, and as it is on the sphere's surface, its radius will be less than or equal to that of the sphere. With these bounds, and taking into account that a sphere is defined by four non-aligned points, we proceed in a similar way to how it was done in a two-dimensional case until we obtain the minimum enclosing sphere

c) Representations for top and tip

To conclude with a sound definition of an exterior representation, we must return to the point left unsolved before, namely, how to cover the bases of the object —top and bottom—. Since both bases are equivalent, we will refer exclusively to the *top* with the understanding that everything stated about the top must also be stated for the bottom.

The top boundary of a translational swept volume is just the closed surface outlined by the edges of its generator convex polygon. As a covering with a set of circles $\{C_i, i=1,...,m\}$ is available for the whole polygon (see the polygon covering algorithm in section 2.1.1), a covering for the top of the solid can be readily obtained by simply replacing each circle C_i by a sphere with the same center and radius.

The covering for the top —and bottom— surfaces of an object, together with the covering for the side surfaces, complete our exterior representation in the sense we initially defined it; i.e., as a total covering for the boundary of the object. However, a new covering must be introduced —in certain cases— in order to preserve those features that are expedient to our purpose, namely, improvability, balance and optimality. A justification for this necessity will be given in the next section when dealing with the quality of a representation.

Instead of covering the sides of an object with a single system of translates $K+\mathbf{a}_j$, we will partition the side boundary into three portions and independently cover each of these portions. What we intend is to give a special treatment to the two side regions adjacent to the bases of the solid. The partition is made by cutting the object through two planes parallel to its bases and *relatively* close to them. In this way two equivalent slices are defined, which will be referred to as the *tips* of the object. One of the tips will be ignored, in the assumption that what is valid for one is also valid for the other.

The final outer representation, then, consists of three different coverings (Fig. 2.5): one for the top —and bottom—, one for the sides —excluding the tips—, and one for the tips. The representation for a tip is computed by application of exactly the same technique already described to obtain the representation for the sides. Naturally, both coverings differ in their defining systems $K+\mathbf{a}_j$. In particular, their set of spheres $\{S_i,\ i=1,...,m\ \}$, their set of circles $\{C_i,\ i=1,...,m\ \}$, their lattice vectors $\mathbf{a}_j$, and m and n are all different.

2.1.3 Quality of a Representation

Three convenient properties for a representation have already been mentioned on several occasions. These are, once more, improvability, balance and optimality. These three concepts take for granted that a certain measure of the quality of a representation can be established. A criterion is clearly necessary to decide on the degree of excellence of a representation; moreover, to state if a representation is balanced, a way of comparing the qualities of different portions of a representation is also needed.

In the mathematical theory of packing and covering the only available parameter to evaluate the quality of a covering is its *density*. For a given covering of space by a system $K+\mathbf{a}_j$, this density

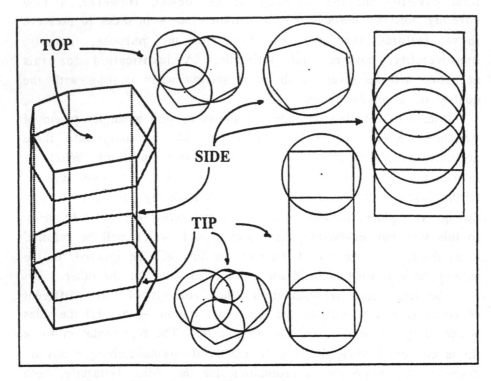

Figure 2.5   Instance of a complete outer representation

$\rho(K+\mathbf{a}_j)$ is described as the limiting ratio of the sum of the measures of those sets of the system of translates, which lie in a large cube, to the measure of the cube, as it becomes infinitely large. It can be shown [Rogers, 1964] that if set K is bounded and has a positive measure, and if the system of translates $K+\mathbf{a}_j$ forms a covering, then

$$\rho(K+\mathbf{a}_j) \geq 1.$$

This definition might be adapted to the problem of our interest, that is, to a covering for the boundary of a solid object. However, density does not comply with our requirements, as it only provides information about how optimal a covering is, in the sense that the greater the density is, the more compact the covering is, and the more number of spheres there are. We are rather concerned with

how well a representation with spheres approximates the *shape* of
the actual geometric object. We want a representation that improves
this approximation upon all its predecessors by better conforming to
the outer form of the object. The underlying idea, upon which all the
following definitions rest, is that a plausible measure of the quality
of a representation can be obtained from the measure of the error
set E. This set is composed of those points Q that belong to a certain
sphere of the covering but are not contained inside any real object.
That is,

$Q \in E$ if and only if $\exists S_j \in S \mid Q \in S_j$ and $\forall i$ $(i=1, 2, ...)$ $Q \notin O_i$,

or simply, $Q \in S$ and $Q \notin O$, or just, $E = S - O$.

Where:

$O = \{O_i, i=1,..., m\}$ is the set of all objects and

$S = \{S_j, j=1,..., n\}$ is the set of all spheres in the representation.

The intuitive idea that supports this definition of E is that this
error set characterizes the discrepancy between the representation
and the real objects, in such a way that a measure of its size serves
to quantify the quality of a representation. It seems evident that the
more points we have in the representation that do not really belong
to the object, the worse the attempted approximation will be. The
most exact representation, therefore, would be one that had an
empty error set.

When considering a choice for a measure of the set E, one may
intuitively think of volume as the best suited to our needs. At second
glance, however, one realizes that volume is lacking a *feeling* about
proximity; two representations with the same error volume may
notably differ in their quality: it is evident that in case (a) of figure
2.6, the real form of the objects is much worse approximated than in
case (b), in spite of the fact that for both cases the error set has a
similar volume. The reason for this apparent paradox is that even
though the sizes of the figures are comparable, their shapes are
radically different, which makes their representation —by a single
circle in this case— very unequal in quality.

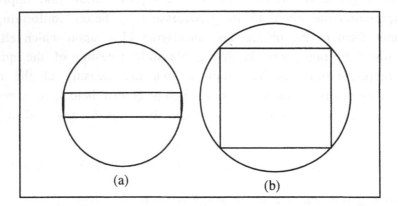

(a)   (b)

Figure 2.6   The volume as a measure of error

Instead, a coefficient —called δ— with the dimensions of length has been introduced. The δ coefficient for a covering may be described as the ratio of the volume of its error set to the surface of the covered boundary of the object:

$$\delta = \text{Vol } (E_i) \text{ / Surf } (BO_i),$$

where E_i is the error set corresponding to object BO_i, and BO_i is its outer boundary. This definition corresponds to the average error distance when we approximate the exterior boundary of the true object by the boundary of the representation.

Furthermore, the coefficient δ offers an additional advantage which is that the previous definition of δ can be directly extended to a partial covering of an object. We have already stated that our representation method is based on a partition of the object into 5 portions; a partition is defined independently for each one. Then, if we consider each of them as a *partial covering* of the object, we can apply to them the concept of δ to obtain a measure of each partial covering. Furthermore, if we associate any portion of the surface of a solid with a set of spheres that covers it, we can likewise define the error set of this *local covering* as a coefficient δ which will give us an idea of its quality. This capacity to define partial qualities will become essential when constructing balanced representations since, by doing so, we will be able to decide which parts of the representation contribute to a greater extent in making it worse.

One last question that is worth being considered is the reason why we have not defined a measure with no dimensions that, like density ρ, does not depend on the real dimensions of the problem. Indeed, we could argue that as δ would be expressed in units of distance, it would be an absolute measure and not a relative one, so that a larger object will be associated with a greater δ value than that of a smaller object regardless of how their representations adapt to the real shape of the objects. This argument is rigorously true and valid; nevertheless, what is presented as a possible drawback really is not that at all, and not only has it been taken into account, but, moreover, it is a desired effect.

It is not our intention to construct an abstract representation of an isolated object, rather, the precision with which an object is represented will be conditioned by its relationships with the rest of its surrounding objects. If we go back for a moment to our main problem —motion planning— we will realize that what is important are the real absolute distances between real objects —robots and obstacles— and these will be the distances that will advise us about how to improve the representation of a certain object at a given moment. In any case, if we wish to know to what extent a certain object adapts itself to our representation method without taking into account its dimensions, a relative adimensional parameter could be defined by merely dividing δ by a distance that gives an idea of the size of the object in question; for instance, this distance could be taken as the diameter of the smallest enclosing sphere (in the first outer representation).

In the present approach several quality coefficients have been introduced to characterize the various coverings making up a representation since, as we have mentioned, the above definition for δ can be applied to a total as well as to a partial covering without modification.

a) Errors in the planar case

Obviously, the above definition of the error set E, for a covering with spheres, can be extended to any covering disregarding the nature of the geometric objects in its generating set. Then, the previous definition for δ can be extended to the planar case if we take surface as a measure of the error set and divide by the length of the covered boundary. This parameter will be called ε:

$$\varepsilon = \text{Surf } (E_i) \text{ / Length } (BO_i),$$

for a given planar object O_i, its boundary BO_i, and error set E_i for a certain representation.

A local ε quality coefficient has been introduced to measure the quality of a partial covering by just one circle for a portion of a polygon boundary (Fig. 2.7):

$$local \; \varepsilon = F \text{ / length of list } (A \; B \; C \; D),$$

where F is the surface shown in the figure, and the length of the list of points is just: $AB + BC + CD$.

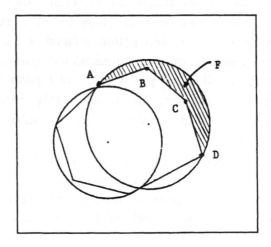

Figure 2.7   Local ε quality coefficient

For a complete polygon, ε coefficient is similarly computed (Fig. 2.8) by adding the different error surfaces and dividing by the polygon perimeter. As we will use this measure both for the side representation and the tip representation, we will use later on ε_s and ε_{ti}, respectively.

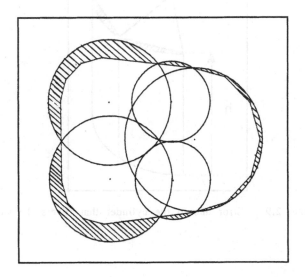

Figure 2.8 ε coefficient for a complete polygon

b) Side and tip errors

For a three-dimensional object, we have already defined a covering for its side surface. To quantify its quality, let us consider first the computation of the local δ coefficient for a set of spheres belonging to a single system of translates $S_i + a_j$ ($j=1, 2, ..., n$) corresponding to a cylinder having C_i as generating circle. Two different parameters are introduced: first one that measures the error due to the cylinder that covers the sides (Fig. 2.9), and second one for the error due to the spheres of system $S_i + a_j$ that cover the cylinder (Fig. 2.10).

The former error is quantified by means of a local quality coefficient. This coefficient is defined as the ratio of the error set of the cylinders (see Fig. 2.9) to the lateral surface of the solid. It can be easily shown that this coefficient is equal to ε coefficient for C_i. In this particular case adding one dimension amounts to adding a factor h that cancels out in the definition of δ to yield ε. For this reason it will be called *side* ε or ε_s.

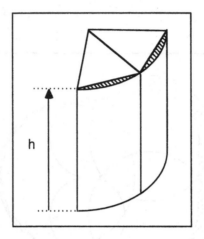

Figure 2.9   Error due to the cylinder that covers the sides

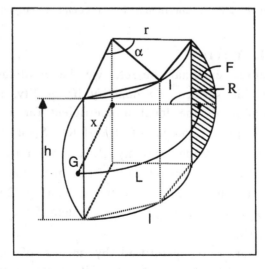

Figure 2.10   Error due to the spheres in $S_i + a_j$ that cover a cylinder

Another local parameter —called *side* δ or δ_s— is introduced to account for the latter error, namely, that of the covering spheres with respect to the cylinder they cover. It can be described, similarly, as the ratio of the error set of the spheres (considered as a covering for the cylinders) to the lateral surface of the covered cylinders (see Fig. 2.10):

$$side\ \delta = F\ L\ /\ h\ l = F\ x\ \alpha\ /\ h\ r\ \alpha = (R^2 - r^2)\ /\ 3r,$$

where F is the surface shown in the figure, L is the length of the curve described by its center of gravity G, l is the length of the shown circle segment, R is the radius of the spheres S_i, and r is the radius of C_i. Obviously, this value of local δ does not change if we take into account all spheres in the system $S_i + a_j$ ($j=1, 2, ..., n$).

For the whole covering for the sides, *side* ε turns out to be equivalent to ε coefficient for the planar case, while *side* δ is given by:

$$side\ \delta\ = \Sigma\ F_i\ L_i\ /\ \Sigma\ h_i\ l_i =$$
$$= 1/3\ \Sigma\ \alpha_i\ (R_i^2 - r_i^2)^{3/2}\ /\ \Sigma\ \alpha_i\ r_i\ (R_i^2 - r_i^2)^{1/2},$$

where both sums extend to all the systems in the set $\{S_i + a_j, i=1, 2, ..., m\}$.

It has to be noted that all these coefficients can be defined, in exactly the same manner, for the tip covering. Moreover, these coefficients may also be used to characterize different portions of a representation. In this way we can discern the contribution of the components of a representation to its final total quality. We will use δ_s for the case of the side represenation and δ_{ti} for the tip representation.

c) Top error

The top covering has an associated parameter —denominated *top* δ or δ_{to}— that provides a measure for the quality of the top portion of a representation. It is described in a similar fashion following the general definition of δ. This *top* δ may, however, lead to an invalid result in certain cases. Namely, when the covering for the side or tip partially surpasses the top covering, as an unwanted collateral effect. If this is the case, the actual error for the top is due to the portion of the side or tip representations covering the top, and not to the representation explicitly computed for the top. To deal with this situation, a couple of new local coefficients have been introduced; they will be referred to as *false side* δ and *false tip* δ —or δ_{fs} and δ_{ft}, respectively—, their meaning being obvious given the general definition of δ when only those spheres that partially cover the top of the object are taken into account.

The previous effect justifies the necessity of the introduction of a specific representation for the tips as it can be understood from Fig. 2.11. Improving the covering for the sides by using more spheres implies that the covering for the top becomes worse. Actually, the better the first one is, the worse the second is. If no tips are defined, the only possible solution for combating this paradoxical effect is adding more spheres to the generating set K for the system $K+a_j$, but this would produce a fast increment in the total number of spheres in the representation. Instead, if special coverings for the tip regions are defined, the quality of the

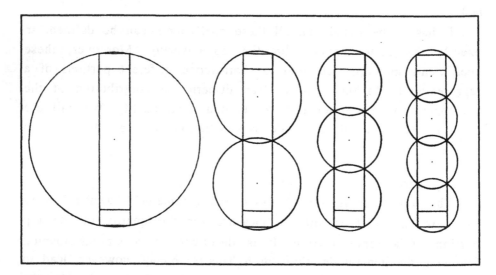

Figure 2.11 Improving the covering for the sides by using more spheres makes worse the covering for the top

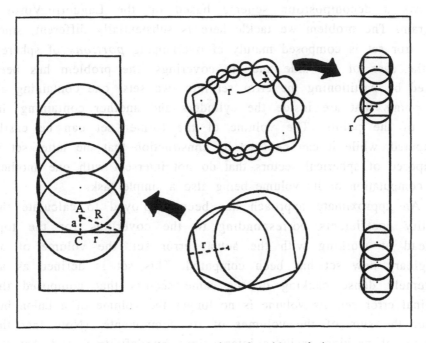

Figure 2.12   Coverings for side and tip

representation is locally improved while still keeping a small number of spheres for the central side region (Fig. 2.12).

d) Total error

Finally, a parameter is needed to evaluate the quality of a given representation taken as a whole. For this purpose, a coefficient called *total* δ or δ_{tot} is defined as the ratio of the error set for all the spheres in the representation to the entire external surface of the object. This parameter will be used later to verify the convergence of a sequence of representations towards a zero-error limit.

A remark has to be made now with respect to the computations involved in the previous definitions. All of them are simple enough with the exception of the problem of calculating the volume of a union of spherical sectors. An $O(n^2)$ algorithm has been proposed [Avis, Bhattacharya and Imai, 1988] for the case of computing the volume of a set of complete spheres, this method follows a decomposition scheme based on the Laguerre-Voronoi diagram. The problem we tackle here is substantially different, since our error set is composed mainly of overlapping *portions* of spheres. In the case of the side and tip coverings, the problem has been solved by partitioning the error set into two sets: one containing all the points that are inside the cylinders and another containing the rest of the .points. The volume of the former set can be easily computed, while it can be seen by construction that the latter set is composed of spherical sectors that do not intersect with one another, the computation of its volume being also a simple task.

An approximate approach has been employed to calculate the quality coefficients corresponding to the coverings for the top. Instead of working with the actual error set, the volume of an imaginary new set has been computed. This set is defined as an extremely dense packing of the same sectors that composed the original error set. Its volume is no longer the volume of a union but rather the sum of the volumes of its components. (Note that the concept of packing excludes intersections by definition, and that the density of this packing is ideally taken as unity to neglect interstitial spaces). Then, the covered surface is accordingly enlarged, in a similar fashion, to preserve the value of the resulting δ coefficient. This technique is valid because density is of no concern for the sake of computing a δ coefficient: remember that an average distance is not affected by proportionally increasing both the volume and the surface that define it.

Finally, the exact volume over the top is needed, since it contributes to the volume of the total error set. To compute an estimate of this value, the volume of the imaginary packing has been normalized, as it were, by means of a factor defined as the ratio of the actual surface of the base polygon to the enlarged surface associated with the packing.

2.1.4 The Expert Spherizer

Now we stop to consider the question of how a given exterior representation can be improved to result in a more accurate, more balanced representation, and keeping, at the same time, the number of involved spheres as low as possible. We will also return to some issues left unsolved before (see section 2.1.1) regarding the heuristics used for the partition of polygons.

The present approach follows a sequential track in the sense that each new representation improves upon the preceding one. In this manner, a representation is obtained by modifying its immediate predecessor by means of performing a certain action on it. The effect of this action usually amounts to substituting a certain set of spheres with a more numerous group of new spheres that refine the portion of the representation assigned to the original set and, consequently, giving the result of an improved complete representation.

To summarize, the current problem can be stated as follows: taking as input a certain representation characterized by a set of quality coefficients, choose among all possible actions the one that will modify it to yield the best resulting representation. The proposed solution is based on a heuristic system —named *the Expert Spherizer*— that is in charge of deciding which action is the most adequate for a given situation. This *expert* has been constructed with the structure of a rule-based system.

a) Two-dimensional case

In section 2.1.1 we presented two techniques to compute coverings with disks for the boundary of a polygon and for its entire surface. Both methods were based on a partition and a farther subdivision after selecting the *worst* portion of the partition. Let us discuss now the details of this process.

In the first case, the boundary of the polygon is first divided into *efficient edges*, instead of just its edges. If the angle between an edge and its neighbor edge is greater than a certain angle —135° in our case—, these edges are considered to be part of the same efficient edge. To justify this decision we must consider that when assigning an error surface to a portion of a polygon in 2D, a sensible choice would seem to be to assign such an error surface to each edge of the object. However, a different criterion has been introduced by defining an *efficient edge* as one composed of several contiguous edges of a polygon. The crucial question is the following: if what we intend is a well-suited object model for AI applications (motion planning or spatial reasoning), then what sense does it make to deal with vertices, edges and facets in the geometric model for a solid? It is obvious that no human being takes care of each facet when reasoning about manipulating an object, but rather a general model of the object is used.

Most solid modelers used in CAD systems can only visualize polyhedra, regardless of the particular technique they may have used (CSG, boundary representation, etc.). As techniques aiming at visualizing solids from general surfaces (as defined by its equations for instance) are still at an early stage, usually any object must be approximated by a polyhedron before being visualized. Perhaps this fact has influenced the models used in motion planning, for most of them always test all possible collisions for all the involved vertices, edges and facets [Boyse, 1979], [Lozano-Pérez, 1987], [Canny, 1987], [Donald, 1987]. In an actual complex design this may lead to lengthy, unnecessary computations. If we consider the amount of such elements that may appear in a real detailed design, we will realize that a great deal of time will be spent in making verifications on unnecessary or unrelated collisions. What spatial reasoning really requires is the use of descriptions based on the notion of *hierarchy of detail*. This amounts to using a multiple representation for each object, in such a way that we will have very simple or very complex representation in this hierarchy to be used in each case according to our needs.

Let us consider the example in Fig. 2.13. Polygons (a) and (b) have very similar shapes, but while (a) is a triangle, (b) has 11 edges. There is no straightforward relationship between the shape and the number of edges or vertices. While it seems a good criterion to assign an error zone to each edge of the triangle (Fig. 2.13(c)), what would be the sense of assigning an error zone to each one of the 11 edges of Fig. 2.13(d)? A higher level structure, the *efficient edge*, has to be introduced to represent the notion of *lateral zone* of a polygon, independently of its vertices. In this way, Fig. 2.13(e) would be composed of only three efficient edges.

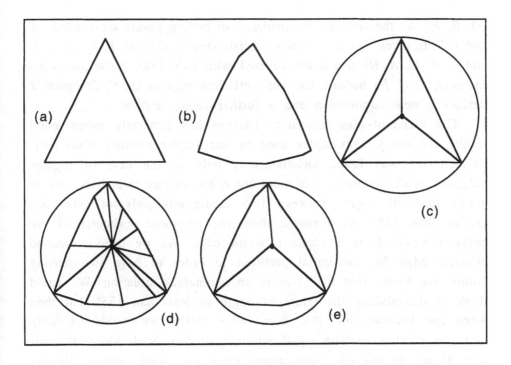

Figure 2.13   An instance of the efficient edge heuristics

This issue will become extremely important when dealing with generalized polygons having curved boundaries. These curves will be approximated by a set of edges and vertices. As we will see in chapter 5, the efficient edge heuristics will make possible an straighforward extension of the model to these generalized polygons.

A group of efficient edges defines a *list*, each list being covered with just one disk. For every efficient edge there is a portion of that disk associated with it, in the sense that this portion covers, so to speak, the efficient edge. Given a certain covering for the whole polygon, the worst list is defined as the list that contains the worst efficient edge. This edge, in turn, is the one whose associated portion of disk has the largest surface. Then, to subdivide the selected list, we cut it through the point on the worst efficient edge that is defined by half the angle subtended by the edge from the center of its covering circle. For instance, in Fig. 2.14 there is only one list, and the worst two efficient edges are number 1 (*G, A, B*) and number 3

(*D*, *E*, *F*). As the first list is circular, two cutting points are needed: *H* and *I*. This gives rise to two new sublists (Fig. 2.15): (*H*, *B*, *C*, *D*, *E*, *I*) and (*I*, *F*, *G*, *A*, *H*) which are covered with two disks. Now the worst list is (*I*, *F*, *G*, *H*) because the worst efficient edge is (*B*, *C*, *D*); point *J* defines a new subdivision and a further representation.

The efficient-edge heuristic criterion has generally given good results for the test examples used in our implementation. Curiously, its behavior was found unsatisfactory only in the case of regular polygons with a number *n* of sides for $n \geq 8$. In this case, it is easy to verify that all angles between two contiguous polygon sides are greater than 135°. As a result, then, the complete perimeter of the polygon would form a single efficient edge. As we need a second efficient edge for the initial partition in order to have two cutting points, we would find ourselves in an anomalous situation. We could think of diminishing the threshold angle to less than 135°, but then when that becomes less than $\beta = (180 - 360/n)$ we would suddenly move to a situation with equal efficient *n* sides with identical error sets. If we do not take precautions, these sets would appear slightly different due to numerical errors, with absurd partitions and

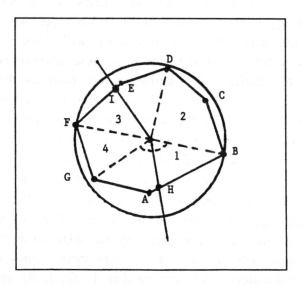

Figure 2.14   Selecting the first two cutting points

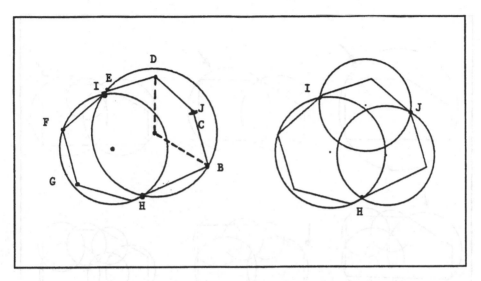

Figure 2.15   New selection and subdivision

representations as a result. Therefore, this case must be handled by means of mere symmetry criteria in order to carry out the partitions. In any case, regular multi-sided polygons are the figures that best adapt to a representation with a single circle, thus making the use of more circles generally unnecessary; furthermore, it is very difficult to improve the initial representation unless a considerable number of circles is used.

A similar scheme is followed in the polygon covering algorithm, with, naturally, certain distinctions. In this case, the worst circle is the one with the largest radius (as this covering is only used in its extension to 3D for the top covering, the worst circle is the one that will give rise to the worst sphere), and each subpolygon is subdivided by means of a segment defined by two points. These two points are located on the two worst efficient edges that are covered by the worst circles (Fig. 2.16).

b) Three-dimensional case

In the three-dimensional case the situation becomes more complex. Starting from a unique outer sphere the sequence of coverings that correspond to each representation must be generated.

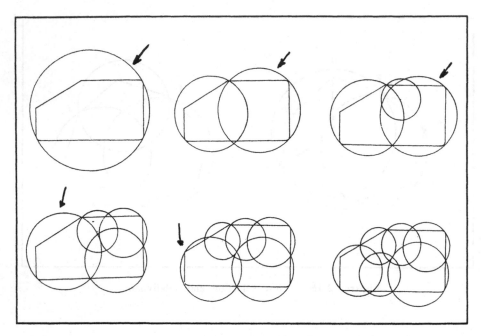

Figure 2.16   Subdivision process in the polygon covering algorithm

We have seen that a general representation for an object is built using five different coverings which are themselves 3D approximations. This fact yields five different regions that can be selected for modification; in addition, each one can be changed in several possible ways. Consequently, the Expert Spherizer is in charge of the decision process involved in the definition of the sequence of representations.

To summarize, we have the following coverings:

(1) The covering for the side region is expressed as the system of translates:

$$\{K_s + \mathbf{a}^s_j; j=1, 2, ..., n_s\},$$

K_s is the set of spheres:

$$K_s = \{S^s_i, i=1,..., m_s\}.$$

Vector $\mathbf{a}^s_j$ is defined as

$$\mathbf{a}^s_j = (2j-1)\mathbf{a}_s + h_{ti}\mathbf{u}, \quad \text{being } \mathbf{a}_s = h_s/2n_s \, \mathbf{u},$$

where $\mathbf{u}$ is a unit vector with the direction of segment H (generator of the object by translational sweeping), and h_{ti} and h_s are the heights of the tip and side boundary.

Moreover, the set of circles

$$\{C^s_i, \, i=1,..., \, m_s\},$$

is the covering for the boundary of the base polygon that has served to build up the 3D covering.

N_s is the total number of spheres in the covering for the sides with system $\{K_s + \mathbf{a}^s_j\}$. Obviously:

$$N_s = m_s \times n_s$$

(2) Similarly, we have the equivalent definitions for the tip coverings:

$$\{K_{ti} + \mathbf{a}^{ti}_j; \, j=1, 2, \, ..., \, n_{ti}\}, \, K_{ti} = \{S^{ti}_i, \, i=1,..., \, m_{ti}\}.$$

Vector $\mathbf{a}^{ti}_j$ is different for both tips; for the lower tip it is

$$\mathbf{a}^{ti}_j = (2j-1)\mathbf{a}_{ti}, \quad \text{being } \mathbf{a}_{ti} = h_{ti}/2n_{ti} \, \mathbf{u},$$

and for the upper tip it is

$$\mathbf{a}^{ti}_j = (2j-1)\mathbf{a}_{ti} + (h_{ti} + h_s)\mathbf{u}.$$

And the set of circles is

$$\{C^{ti}_i, \, i=1,..., \, m_{ti}\}.$$

N_{ti} is the total number of spheres in the covering for one tip with system $\{K_{ti} + \mathbf{a}^{ti}_j\}$. Obviously:

$$N_{ti} = m_{ti} \times n_{ti}$$

(3) Finally, for both base regions the covering system is

$$\{K_{to} + \mathbf{a}_{to}\},$$

Since in this case $n_{to} = 1$.

$$K_{to} = \{S^{to}_i, \ i=1,...,m_{to}\}$$

Vector $\mathbf{a}_{to}$ is $\mathbf{a}_{to}=0$ for the lower base, and for the upper it is:

$$\mathbf{a}_{to} = (2h_e + h_l)\mathbf{u}.$$

The set of circles is

$$\{C^{to}_i, \ i=1,..., \ m_{to}\}.$$

N_{to} is the total number of spheres in the covering for the top with system $\{K_{to} + \mathbf{a}^{to}_j\}$. In this case

$$N_{to} = m_{to}$$

N is the total number of spheres for a certain exterior representation and it is given by:

$$N = N_s + 2N_{ti} + 2N_{to}$$

See Fig. 2.5 for an example showing the different sets of spheres that compose the representation.

Let us now briefly review the set of possible actions among which our expert can choose to improve a given representation. It can decide to start defining a covering for the top, if it detects that the coverings for the side and tip do not guarantee that the top is covered. If the subdivision between sides and tips has not been established, it can determine the convenience of doing so. It can also decide to add more cylinders to cover the sides or the tips, or to add more spheres to cover top and bottom. And finally, the expert can append more spheres to cover the different cylinders already defined. The basic possible choices can be, then, summarized as follows (obviously, each of them is later subdivided in different ways):

1.- *Introduce the coverings for the bases.* In the first representation a single sphere defines K_s, and $N_{ti}=N_{ti}=0$ with $n_s=m_s=1$. In a certain moment during the subsequent process the specific top covering $\{K_{to}+\mathbf{a}_{to}\}$ may be introduced.

2. *Introduce the coverings for the tips.* Similarly, in a certain moment the tip covering $\{K_{ti}+\mathbf{a}^{ti}_j\}$ may be introduced. That amounts to the partition of the outer boundary of the object into three different regions: two tips and one central side; $\{K_s+\mathbf{a}^s_j\}$ will have to be redefined.

3.- *Modify set K_s.* That is, increase the number of spheres in the generating set for the side covering. That is equivalent to the modification of set $\{C^s_i\}$.

4.- *Modify set K_{ti}.* That is, increase the number of spheres in the generating set for the covering of the tips. That is equivalent to the modification of set $\{C^{ti}_i\}$.

5.- *Modify set K_b.* That is, increase the number of spheres in the generating set for the top covering. That is equivalent to the modification of set $\{C^{to}_i\}$.

6.- *Modify sequence $\{\mathbf{a}^s_j\}$.* A new sequence $\{\mathbf{a}^s_j\}$ is defined. The number of vectors composing the system of translates $\{K_s+\mathbf{a}^s_j;$ $j=1, 2, ..., n_s\}$ will be increased.

7.- *Modify sequence $\{\mathbf{a}^{ti}_j\}$.* A new sequence $\{\mathbf{a}^{ti}_j\}$ is defined. The number of vectors composing the system of translates $\{K_{ti}+\mathbf{a}^{ti}_j;$ $j=1, 2, ..., n_{ti}\}$ will be increased.

c) Structure of the Expert Spherizer

The basic premise, upon which the selection of the previous actions relies, is that a balanced and quasi-optimal improvement to a representation can be obtained by detecting the zone with the worst local quality and subsequently refining its representation. It has to be noticed that a certain action can improve the local representation for a zone, while, at the same time, it makes the quality of another zone worse. Several heuristics have been employed to deal with this kind of situation. For this problem we have chosen to equip the Expert Spherizer with the structure of a *rule-based system*.

Rule-based systems [Buchanan and Shortliffe, 1984] or *knowledge-based systems* have been very successful tools in Artificial Intelligence. In fact, these systems have given rise to *expert systems* [Hayes-Roth, Waterman and Lenat, 1983] that may be considered as the first applications of Artificial Intelligence to become an incoming-producing commercial product. Basically, its effectiveness lies in the use of a description of an expert's knowledge of a certain domain, which is stored as a set of heuristic rules. These rules form the *knowledge base* and represent the broken-down knowledge that is required to take on a problem which, due to its complexity, cannot be solved by an algorithm. A system called *inference engine* is in charge of *triggering* the rules that may apply for a given particular situation so that by chaining the results of some rules with others, a final conclusion is reached. A detailed description of the implementation of an expert system based on rules constructed in Lisp language can be found in [Huet, 1986], [del Pobil, Muñoz and García, 1988].

The expert spherizer works by following the fundamental criterion of detecting the representation area whose local quality is the worst, in order to later refine its representation by means of a certain action. Of course, it must be taken into account that the applied action may have side effects that must be corrected. The quality coefficients that will serve to make the main decisions are: δ_s and ε_s for the side covering; δ_{ti} and ε_{ti} for that of the tips; δ_{to}, δ_{fs}, δ_{ft} for the top and bottom; and finally, δ_{tot} which characterizes the complete exterior representation.

Next, we will comment on the main heuristic rules that the expert spherizer uses. From the relationships between the previous quality coefficients it will decide which is the best action to be applied.

Rule (0). If the specific top coverings $\{K_{to} + \mathbf{a}_{to}\}$ have not been defined, and it is necessary that they exist to guarantee a correct representation, then the first covering is defined $\{K_{to} + \mathbf{a}_{to}\}$ with $m_{to} = 1$. We must point out two cases

> <u>Rule (0.1)</u>. If the covering $\{K_{ti} + \mathbf{a}^{ti}{}_j\}$ has not yet been defined, verify if $\{K_s + \mathbf{a}^s{}_j\}$ guarantees that the top and bottom are covered; if this is not so then define $\{K_{to} + \mathbf{a}_{to}\}$.

Rule (0.2). If the covering $\{K_{ti}+\mathbf{a}^{ti}{}_j\}$ has already been defined, verify if it guarantees that the top and bottom are covered; if this is not so then define $\{K_{to}+\mathbf{a}_{to}\}$.

Rule (1). When the covering $\{K_{ti}+\mathbf{a}^{ti}{}_j\}$ exists and it contributes to the top and bottom, it may be necessary to correct it in certain cases. We must distinguish between two cases.

Rule (1.1). If $\{K_{to}+\mathbf{a}_{to}\}$ has been defined, verify if $\{K_{ti}+\mathbf{a}^{ti}{}_j\}$ surpasses it, that is, $\delta_{ft}>\delta_{to}$. This rule tries to avoid anomalous situations in which the tip covering, whose mission is to cover part of the side boundary, introduces an error by exceeding the top and bottom coverings. Then, the set K_{ti}, must be modified, so in this way, upon decreasing the radius of the spheres of the generating set, its contribution to the error set on the top and bottom will be less.

Rule (1.2). If $\{K_{to}+\mathbf{a}_{to}\}$ does not exist, in this case $\{K_{ti}+\mathbf{a}^{ti}{}_j\}$ is in charge of covering the top and bottom. If its error set on the top and bottom generates an average error distance $\delta_{ft}>\delta_s+\varepsilon_s$, that is, worse than the one due to the side covering, then it will be necessary to correct K_{ti} by introducing more spheres.

Rule (2). This rule considers the effect produced when the side covering $\{K_s+\mathbf{a}^s{}_j\}$ in some way affects the top and bottom. Logically, this case will come up only if the specific tip covering has not been defined; otherwise $\{K_s+\mathbf{a}^s{}_j\}$ could never surpass the top and bottom ($\delta_{fs}=0$). We should consider different cases:

Rule (2.1.1). If $\{K_{to}+\mathbf{a}_{to}\}$ has been defined, verify if $\{K_s+\mathbf{a}^s{}_j\}$ surpasses it, that is, $\delta_{fs}>\delta_{to}$. This rule tries to avoid anomalous situations in which the side covering introduces an error as it exceeds the top and bottom coverings. A first possibility to solve this situation is to introduce an initial $\{K_{ti}+\mathbf{a}^{ti}{}_j\}$ with $m_{ti}=n_{ti}=1$. This solution is not always convenient Whether it is convenient or not to define these coverings depends on a certain heuristic coefficient γ.

Rule (2.1.2). If in the previous situation it is not possible to define the tip covering, the only possibility left to improve δ_{fs} is by modifying K_s so that the spheres that make it up will have smaller radii.

Rule (2.2.1). If $\{K_{to}+\mathbf{a}_{to}\}$ has not been defined, then the top and bottom must be covered by $\{K_s+\mathbf{a}^s_j\}$. If the error set on them is such that $\delta_{fs}>\delta_s+\varepsilon_s$, that is, with a quality coefficient worse than that of the local covering of the side surfaces by $\{K_s+\mathbf{a}^s_j\}$ itself, then it will be necessary to define the tip covering if it is possible.

Rule (2.2.2). If in the last case it is not convenient to define an initial $\{K_{ti}+\mathbf{a}^{ti}_j\}$, it will be necessary to correct K_s by introducing more spheres.

Rule (3). After having considered in the previous rules the side effects of $\{K_s+\mathbf{a}^s_j\}$ and $\{K_{ti}+\mathbf{a}^{ti}_j\}$ on the top and bottom, now we are going to search out which is the worst partial covering among those corresponding to the sides, tips, and top (and bottom), so as to try to correct it. In this rule we look at the less frequent case in which the tip covering is the worst, that is, $\delta_{ti}+\varepsilon_e>\delta_s+\varepsilon_s$ and furthermore $\delta_{ti}+\varepsilon_{ti}>\delta_{to}$. We have two possible actions to improve the two coefficients —δ_{ti} and ε_{ti}— which contribute to the quality of $\{K_{ti}+\mathbf{a}^{ti}_j\}$. We will give precedence to the first of these actions over the second by introducing a coefficient $\kappa<1$ which will multiply ε_{ti}. The reason is that it does not make sense to try to adjust set K_{ti} until the cylinders are well approximated by spheres. Furthermore, the first of these operations always gives good results, while the second is more uncertain. The concrete value used in the implementation was $\kappa=0.5$

Rule (3.1). If $\delta_{ti}>\kappa\varepsilon_{ti}$, we modify the sequence $\{\mathbf{a}^{ti}_j\}$ in order to improve δ_{ti} so that when n_{ti} increases by 1 the cylinder approximations improve.

Rule (3.2). If, on the other hand, $\delta_{ti}\leq\kappa\varepsilon_{ti}$ then, to improve ε_{ti}, we modify the set K_{ti}, in such a way that the error set of the cylinders in relation to the tips of the object decreases.

Rule (4). Let us now consider the case in which the covering $\{K_s + \mathbf{a}^s_j\}$ is the worst, that is, $\delta_s + \varepsilon_s \geq \delta_{ti} + \varepsilon_{ti}$ and furthermore $\delta_s + \varepsilon_s \geq \delta_{to}$. As in rule (3), we have two possible actions to improve this covering, either improve δ_s, or ε_s. Likewise, we will favor δ_s over ε_s, and we will use the κ factor defined in the exact manner as in the previous case. The reason is obvious, since the coverings for the tips and sides are conceptually the same although they are applied to different situations.

    <u>Rule (4.1)</u>. If $\delta_s > \kappa \varepsilon_s$, to improve δ_s we modify the sequence $\{\mathbf{a}^s_j\}$ so that when n_s increases by 1, the approximation of the corresponding cylinders improves.

    <u>Rule (4.2)</u>. If, on the other hand, $\delta_s \leq \kappa \varepsilon_s$ then, to improve ε_s, we modify the set K_s, in such a way that the error set of the cylinders in relation to the tips of the object decreases.

Rule (5). Finally, the last case to be dealt with will be that in which the covering $\{K_{to} + \mathbf{a}_{to}\}$ is the worst, that is, $\delta_{to} > \delta_s + \varepsilon_s$. In this situation we only have one possible action to improve the top and bottom coverings, namely, increasing m_{to} by modifying the set K_{to} (since the sequence $\{\mathbf{a}_{to}\}$ is fixed).

The control strategy used to apply these rules is one of the most commonly used in rule-based systems. The rules take priority according to their order, in such a way that first, we try to trigger the rule (0.1), if it is not applicable, we move to (0.2), and so on until we get to (5). Therefore, the order in which the rules have been stated is not arbitrary; moreover, by their very definition, it is guaranteed that at least one of them is always applicable. When the first valid rule is found, it is *triggered*, that is, it carries out the corresponding action thus creating a new representation with a new set of parameters and quality coefficients that characterize it. Next, it would start again with the first rule to discover which one can now be applied to the new representation.

The spheration process, therefore, follows a constructive problem-solving approach with the step by step generation of new representations which are improvements on the previous ones. Naturally, the starting representation is the one in which the whole object is enclosed in a single sphere, in such a way that only $\{K_s + \mathbf{a}^s{}_j\}$ is defined, being $m_s = n_s = 1$ and $N_{ti} = N_{to} = 0$. The second representation will be obtained from this one by applying one of the rules, and so on. We will leave the discussion of the results of this spheration process for the final section of this chapter which deals with the convergence of the method.

2.2 Interior Representation

As it was stated when introducing the concept of a double spherical representation for an object, a parallel description of this object is made by means of two sets of spheres. The first set is the outer representation that has already been analyzed; it is time to analyize the second set, named the interior representation of an object.

An interior sphere is defined as one that is completely contained inside one of the objects being represented. Namely,

$$\forall P \in S_j \ \exists O_i \in O \mid P \in O_i.$$

Where:

$O = \{O_i, \ i=1,..., \ m \ \}$ is the set of all objects.

$S = \{S_j, \ j=1,..., \ n \ \}$ is the set of all spheres in the representation.

We wish to define a certain set of n inner spheres in such a way that, regarded as an interior representation of the set of m objects O, it can be improved, it is balanced, and it is the best for the given number n of involved spheres, according to some pre-defined criteria. The present definition could be considered a generalization of the concept of a packing with spheres, if the condition of no intersection of the sets of the packing is removed:

$$\cup S_j \subset O, \qquad S_j \cap S_k = C_{jk}.$$

Where the set C_{jk} may be empty or not. The strict mathematical definition of packing compels C_{jk} to be empty, but this would not serve our purpose, for such a packing would not be at all optimal.

We will assume the same restrictions — as in the case of coverings— about the nature of the solid objects to be represented. Similarly, we will use the concept of a system of translates $K+a_j$ ($j=1$, 2, ...) to define one of such generalized packings, as well as the notion of density $\rho(K+a_j)$. Now, with the same constraints as in the covering case, it is

$$\rho(K+a_j) \leq 1.$$

We will now use the notion of density for the generalized packings of a certain object, defining it as the quotient between the measure of the union of the packing sets and the measure of the object. Note that, unlike the case for covering, this definition is now correct, for both the strict-sense case —in which the measure of the union equals the sum of the measures, the intersection being empty— and in the generalized case, since what interests us is to measure which portion of the interior representation of the object is occupied by packing sets.

2.2.1 Planar Case

The problem is now particularized to two-dimensional space. In this section we prove the existence of generalized packings for polygons by means of circles, which are as exact as desired.

a) Voronoi segments and maximal circles

For the previous purpose we need a number of preliminary results and definitions concerning the concept of Voronoi Diagram (VD). This is one of the most useful concepts in computational geometry, and it has already been mentioned several times throughout this book. Originally, the VD construction was proposed for a distribution of points in space in the following way [Voronoï, 1908].

Let us assume that the set of points

$$\mathbf{a}_1, \mathbf{a}_2, \ldots$$

is discrete, and that there exists a positive number R such that for every point $\mathbf{x}$ in space, there is a point $\mathbf{a}_i$ from the previous sequence whose distance $|\mathbf{x}-\mathbf{a}_i|$ from $\mathbf{x}$ is less than R. To each point $\mathbf{a}_i$ we associate the set $\Pi(\mathbf{a}_i)$ of all points $\mathbf{x}$ whose distance from $\mathbf{a}_i$ is equal to the minimum distance to points of $\{\mathbf{a}_j\}$. Then, $\Pi(\mathbf{a}_i)$ is the set of all points $\mathbf{x}$ that satisfy

$$|\mathbf{x}-\mathbf{a}_i| \leq |\mathbf{x}-\mathbf{a}_j| \ (j \neq i).$$

It can be shown that the set $\Pi(\mathbf{a}_i)$, defined in this way, is a closed convex polyhedron. Furthermore, each point in space belongs to a least one of such polyhedra and can only belong to two or more of the polyhedra if it belongs to their boundaries.

In other words, we can say that the Voronoi diagram defined by the sequence $\{\mathbf{a}_j\}$ defines a partition of space into regions $\Pi(\mathbf{a}_i)$ which are associated to each point $\mathbf{a}_i$, in such a way that region $\Pi(\mathbf{a}_i)$ contains all those points in space that are closer to $\mathbf{a}_i$ than to any other point of the sequence $\{\mathbf{a}_j\}$ (See an example of VD for a set of 16 points in Fig. 2.17).

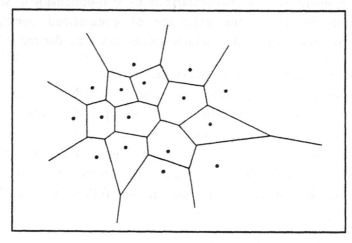

Figure 2.17   Voronoi Diagram for a set of points.

The usefulness of the VD has become evident in its numerous applications. In this way, for the case of n points on the real plane, with usual Euclidean metrics, an algorithm $O(n\log n)$ was found when dealing with the problem of the minimum enclosing circle [Shamos and Hoey, 1975]. Other proximity problems (closest pair, closest neighbor, etc.) can be solved efficiently once the VD for the set of points involved has been constructed [Leven and Sharir, 1987b]. A survey of these problems in the Euclidean case can also be found in [Preparata and Shamos, 1985].

In the last few years, several generalizations of the initial definition of the diagram by Voronoi have been proposed. In this way Edelsbrunner and Seidel [1985] propose a general framework for dealing with VD in such a way that it is considered a partition of a certain domain, which is induced by a finite number of real functions on the domain. Even more interesting is the study on how to define a VD for a set of segments and circles in the Euclidean plane done by Lee and Drysdale [1981]; this variation is sometimes called the generalized Voronoi diagram (GVD). Other possibilities include the use of distance measures that are not usual Euclidean metrics [Canny and Donald, 1988a], or even an *abstract* Voronoi diagram has been proposed in which there is no notion of distance involved whatsoever [Klein, 1988].

In our area of interest —namely, motion planning— approaches based on VD have frequently been presented. As an example, we can cite the theoretical retraction algorithm [O'Dúnlaing and Yap, 1985] which moves the center of a disk following the boundaries of a region defined by the VD of polygonal obstacles; later, a variation of the diagram was used for the case in which a rod moves on the plane [O'Dúnlaing, Sharir and Yap, 1986 1987]. Other examples of handling a moving object —sometimes called a *piano*— in two dimensions are [Nguyen, 1984] as well as [Takahashi and Schilling, 1988], who use the Voronoi diagram as a reference upon which an object is moved, applying heuristic criteria to decide when to carry out rotations. An extension of VD has also been used for the optimal path problem of a point between polyhedral obstacles [Akman, 1987].

For our problem, we are particularly interested in the notion of Generalized Voronoi Diagram (GVD) [Lee and Drysdale, 1981]. For the n edges of a convex polygon, the GVD can de defined as a partition of the polygon into n disjoint regions Π_i; each of these regions is associated with a certain edge L_i of the polygon, in such a way that all the points P that belong to a region share the same nearest edge. That is to say,

$$\Pi_i = \{P; \, d(P, L_i) \le d(P, L_j), j=1,...,n\},$$

Where $d(P, L_j)$ is the euclidean distance from point P to the line supporting edge L_j.

It must be noted that in the general case where the sequence $\{L_j\}$ may represent any set of segments, we could only define $d(P, L_i)$ as the distance between P and its projection Q on the supporting line of L_i, as long as Q belongs to L_i; otherwise, it must be defined as

$$d(P, L_i) = \min(d(P, A), d(P, B))$$

A and B being the endpoints of segment L_i. It is easy to verify —as we will see further ahead— that if segments of the sequence $L_1, L_2,...,$ L_n form the adjacent sides of a convex polygon, the given definition of $d(P, L_i)$ will be sufficient.

Another fundamental difference between the more general case and that of a polygon is that, in the latter, the Voronoi regions will be convex polygons, whereas in the general case some portions of the boundary of a region will be curves. More precisely, the curves corresponding to a segment and the endpoint of another are portions of a parabola (Fig. 2.18). In our case, the common border between two Voronoi regions will be called a *Voronoi segment* V_i, while the common point for three Voronoi edges is named a *Voronoi point* (Fig. 2.19). In the particular case of a convex polygon, all Π_i are also convex polygons.

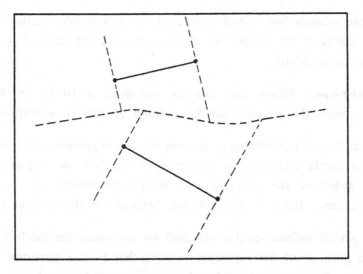

Figure 2.18   Voronoi Diagram for two segments

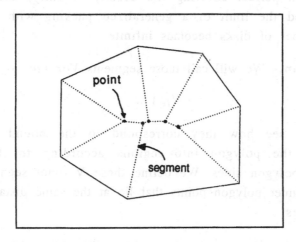

Figure 2.19   Generalized Voronoi Diagram for a convex polygon

The Voronoi diagram defined in this manner has a very important property; namely, that the finite set of Voronoi points is the locus of the centers of all circles that are tangent to at least three edges of the polygon, these circles will be called *maximal circles* C_i. In fact, it is this property that has served to derive an algorithm to construct the GVD, instead of using that of Lee and Drysdale [1981].

Once these centers have been computed, it is a simple matter to find Voronoi segments by connecting them with line segments. Let us see this with some detail.

Theorem.- Given any convex polygon, a finite set of line segments inside the polygon can be defined so that it is verified that:

(1) For each point P belonging to one of the segments, there exists a single circle $C(P)$ with its center at P which is tangent to at least two sides of the polygon, and which, moreover, is completely inside the interior of the polygon (except for the tangent points).

(2) The set of infinite circles obtained by assigning the circle $C(P)$ to each point P of the segments verifies that it is a covering for the polygon by means of circles that are inner to the polygon.

(3) This is a perfect covering (its density is unity) and it can be considered the limit of a generalized packing with disks when the number of disks becomes infinite.

Definition.- We will call those segments *Voronoi segments*

$$V_1, V_2,...$$

and we will see how they correspond to the boundaries of the partition of the polygon into regions according to the Voronoi diagram for polygon sides. We define these Voronoi segments as the locus of all inner polygon points that lie at the same distance from at least two sides.

Properties.- Voronoi segments have the following properties:

(1) All Voronoi segments are a portion of the bisector of two polygon edges.

(2) Two adjacent polygon edges will define a Voronoi segment with origin in their common vertex, which we will call a *terminal segment*

(3) If two edges are not adjacent, and therefore do not intersect each other on a common vertex, their bisector may or may not contain a Voronoi segment.

(4) Voronoi segments intersect one another at points where at least *three* segments coincide. A property of these points is that they are the center of circumferences tangent to at least *three* polygon edges

Algorithm.- To construct Voronoi segments, we will use an algorithm based on the following lines of reasoning:

Given a convex polygon defined by its p edges

$$L_1, L_2,..., L_p$$

1.- For all possible groupings of three edges, the circumference which is tangent to those three edges is found. Its center is the intersection point of the three bisectors for the three pairs of edges. (We consider it unnecessary to show that the three bisectors intersect one another at the same point.)

2.- We verify which of these circumferences are inside the polygon and we discard those that are not. This first resulting set of interior circles will be called *set of maximal circles*

$$C^m = \{C_i \; ; \; i=1, 2,..., m_m\}$$

and will serve to define the Voronoi segments; indeed, the centers of these circles coincide with the intersections between the segments.

3.- As each center of a maximal circle is associated to three bisectors that define it, and as a segment is always on a bisector, then each center of a maximal circle corresponds to the intersection of three Voronoi segments.

4.- A bisector cannot contain more than one segment, that is, there cannot be more than one segment on a bisector. Indeed, we have two possible situations:

a) If the bisector starts at a vertex of the polygon, the segment limited by a maximal center indicates when circles tangent to both edges L_1 and L_2 cannot become greater as they are limited by a third edge L_3 (Fig. 2.20).

b) If the bisector corresponds to two non-contiguous edges, its origin will be at an intersecting point Q which is not a vertex of the polygon, then, the circumferences tangent to both edges will be greater the farther its center is away from the aforementioned origin. The set of these circumferences will have two limits, one for the smaller circumference and one for the greater (Fig. 2.21). The convexity of the polygon guarantees that only interior circles with centers at a continuous interval of the bisector can exist.

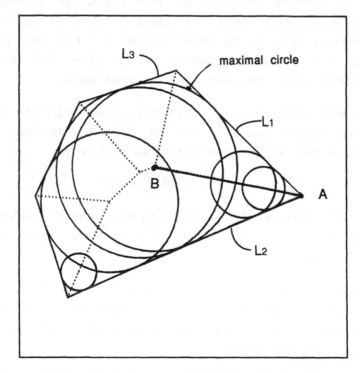

Figure 2.20   Terminal segment limited by point B and vertex A

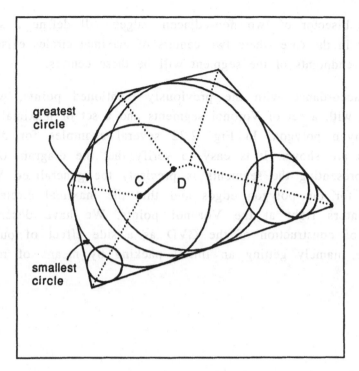

Figure 2.21   A non-terminal segment with two Voronoi
points *C* and *D* as its ends

5.- According to the last result, the bisector that corresponds to non-
adjacent edges will have either two maximal circle centers or it
will have none. If it has two centers, they will define a non-
terminal segment limited by them. A limit case may also arise
where both centers coincide at one single point; or pathological
cases for regular polygons with many edges where several
maximal circles are the same and have the same center, then all
of them should be computed as a sole maximal circle.

6.- The bisector of two adjacent polygon edges will give rise to a
terminal segment whose endpoints are the common vertex of
both edges and the center of a maximal circle associated to the
bisector.

7.- The bisector of two non-adjacent edges will define a segment only in the case where two centers of maximal circles exist on it. The endpoints of the segment will be these centers.

In accordance with the previously mentioned points, we will conclude with a set of Voronoi segments and a set of maximal circles for a given polygon. In Fig. 2.22 several examples for different polygons are shown. It is easy to verify that the diagram obtained from representing the segments is precisely the generalized Voronoi diagram for the polygon edges and that the maximal circles have their centers right at the Voronoi points. We have described a method of construction of the GVD as a side effect of our main objective, namely getting an initial packing by means of maximal circles.

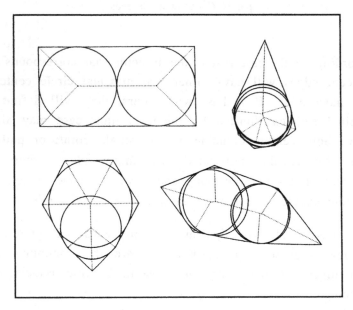

Figure 2.22   Examples of Voronoi diagrams and maximal circles

Other important results that can be deduced from the previous discussion are:

1.- There cannot exist any interior circle inside the polygon that is greater than the circle with the greatest radius of the set $C^m = \{C_i ; i=1, 2,..., m_m\}$.

2.- For a given segment V_i, the elements of the set of circles with centers on the segment and tangent to the edges that define the segment

$$\{C(P); P \in V_i\}$$

have a lower bound and a higher bound —taking the radius as the measure of the size of a circle— which are given by

$$C(A_i) \text{ and } C(B_i)$$

where A_i and B_i are the endpoints of segment V_i. When the segment is terminal, A_i will coincide with one vertex of the polygon and we will take the radius of the circle $C(A_i)$ as null.

3.- A first interior representation is obtained by the generalized covering made up of the set of maximal circles C^m. This representation shows an interesting feature: it *fills up* the inside of the polygon quite well, as all of the circles that make it up are tangent to at least three edges.

b) A perfect covering

We are now in a position to state our main result for the plane. If a disk $C(P)$ is assigned to every point P on a Voronoi edge, so that its center lies on this point, and it is tangent to two edges of the polygon; then a *perfect covering* for the polygon by means of interior circles is defined by the complete set C_∞ of all the disks assigned to the points of all the Voronoi edges (Fig. 2.23).

$$C_\infty = \{C(P); P \in V_i, i=1, 2, ..., v\}.$$

where v is the number of Voronoi segments.

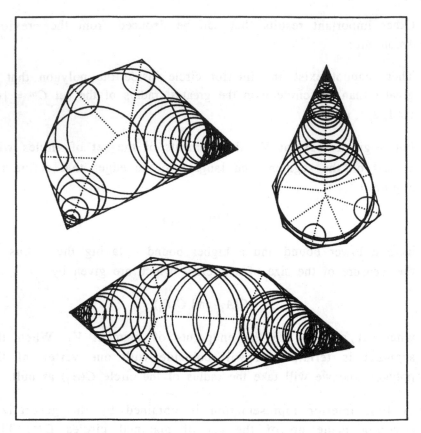

Figure 2.23   Three instances of nearly perfect packings

C_∞ is obviously a covering for the polygon since every point of the polygon is contained in at least one of the circles. Moreover, its density is unity, being at the same time a packing and a covering. Note that a generalized packing becomes also a covering when it reaches its limit of perfection (its density is unity); and that the converse can also be stated for a perfect covering. Indeed, if we define set C_∞ as a sequence $C_\infty = C_1, C_2, \ldots$ and its subset $C(m)$ as

$$C(m) = \{C_1, C_2, \ldots, C_m\};$$

then, a valid definition for C_∞ is

$$C_\infty = \lim_{m \to \infty} C(m),$$

its density being

$$\rho(C_\infty) = \lim{}_{m\to\infty} \rho(C(m)) = 1.$$

(Note that for a covering $\rho(K+a_j)\geq1$, while for a packing $\rho(K+a_j)\leq1$).

c) Building the interior representation

The perfect packing/covering so defined, however, is of no use for practical purposes, since it is composed of an infinite number of circles. A heuristic procedure is called for to select, among this infinite set, the m disks that will best represent the polygon. A constructive scheme is again proposed (Fig. 2.24)

Now our problem can be stated as follows. An enumeration for the elements of C_∞ is needed:

$$C_1, C_2, ..., C_m, ...,$$

in such a way that packing $C(m)$, composed of the first m elements $C_1, C_2, ..., C_m$ of C_∞, becomes an interior representation using m circles

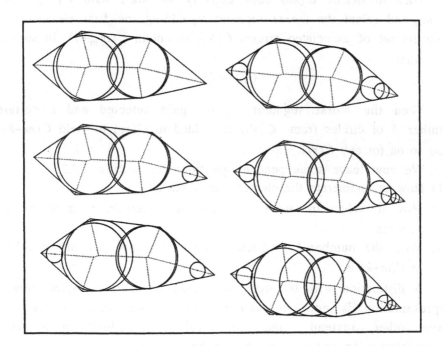

Figure 2.24   Building the interior representation

that is balanced and optimal for the used number of circles, and that can be improved by adding the subsequent elements $C_{m+1}, C_{m+2}, ...$

The first elements of C_∞ will be taken from the set of maximal circles $C^m = \{C_i ; i=1, 2, ..., m_m\}$, so that $C(m_m)=C^m$. Moreover

$$r_i \geq r_j, \quad i<j \leq m_m,$$

where r_i is the radius of C_i.

Now, for $m > m_m$ if we call $C(V_i)$ the subset of C_∞ composed of those circles having its center on V_i, excluding maximal circles, then

$$C_\infty = C^m \cup \{C(V_i); i=1, 2, ..., v\}.$$

And the infinite elements of $C(V_i)$ are denoted as

$$C_{i1}, C_{i2}, ...,$$

in such a way that the elements in each one of these sequences are *conveniently* sorted.

Then to define $C(m)$ constructively we start with $C(m_m)$ and $m = m_m$ and select the *worst* segment V_i adding the first element C_{i1} from its set of associated circles $C(V_i)$ to obtain $C(m_m+1)$, in such a way that

$$C_{m+1} = C_{i1}.$$

Then the worst segment V_i is again selected and a certain number k of circles from $C(V_i)$ are added to $C(m)$ to yield $C(m+k)$. And so on for any $C(m)$.

We now leave some pendant questions for section 2.2.3:
(1) How to enumerate the elements in a set $C(V_i)$.
(2) How to select the *worst* segment in a certain moment in the process.
(3) What the number k of circles from $C(V_i)$ that become members of $C(m+k)$ is.

A distinction with the outer representation must be noted: now a representation is not improved by substituting some circles with some other; instead some new circles are added to a given representation in order to produce a better one.

2.2.2   Extension to Three-dimensional Space

The technique that leads us from the plane to the 3D space is rather similar to the approach presented for the exterior representation. Let us point out its main features as well as some distinctions with the exterior case.

To begin with, a cylinder is generated by each circle C_i belonging to the generalized packing for the generator polygon of the solid object. Then, for *each* such cylinder a generalized packing with spheres $S_i + a_{ij}$ ($j=1, 2, ..., n_i$) is defined (Fig. 2.25). S_i shares its radius r_i with C_i, and its center is located at a distance r_i from the center of C_i along the generating segment H of the cylinders; n_i is the number of spheres in the packing for the cylinder and can be defined as:

$$n_i = \text{excess} \left((h - 2r_i)/(\lambda r_i) \right) + 1.$$

Where h is the length of H, and the function *excess* performs an excess rounding. The sequence of vectors $\{a_{ij}\}$ is defined as:

$$a_{ij} = j a_i, \quad (j=1, 2, ..., n_i)$$

where a_i is a vector with the direction of the equivalent vector for

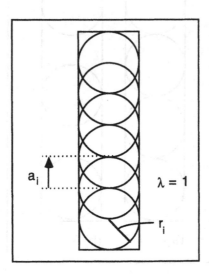

Figure   2.25   A generalized packing with spheres for a cylinder

the exterior case, and whose magnitude is given by:

$$|\mathbf{a}_i| = (h - 2r_i)/(n_i - 1)$$

The density of the packing is governed by the coefficient λ ($0 \le \lambda \le 2$). Note that according to the previous definition for n_i, the number of spheres per cylinder becomes greater as λ is made smaller. If λ equals unity, the distance between the centers of two neighbor spheres is approximately equal to their radius (Fig. 2.25); for $\lambda = 2$ this distance would be around their diameter, for $\lambda > 2$ the spheres are too spread out to make an acceptable packing (Fig. 2.26), and when λ tends towards zero the number of spheres becomes infinite with their centers at an infinitesimal distance.

As a final corollary, we can establish our main result to the effect that the set of systems

$$\{S_i + \mathbf{a}_{ij}, j=1, 2, ..., n_i, i=1, 2, ..., m\}$$

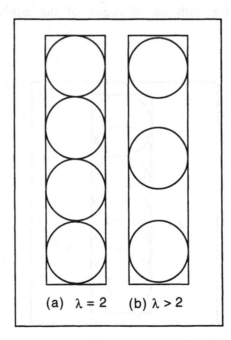

(a) $\lambda = 2$ (b) $\lambda > 2$

Figura 2.26   Influence of λ on the packing density

constitutes a generalized packing for a translational sweeping object, using only spheres (m is the number of circles in the packing for the generating polygon of the solid). Note that now we do not define a single system of translates, as we did in the exterior case, rather we define a set of systems. The reason for this is that the sequence of vectors $\{a_{ij}\}$ is different for each cylinder. We considered this possibility for the exterior representation, but, in that case, we opted to define a single vector a common to all of the cylinders, in such a way that each one of them was covered with the same number n of spheres. In the interior case we have preferred to manage the number of spheres n_i for each cylinder by means of a factor λ, so that the worst admissible density ($\lambda=2$) practically corresponds to a strict cylinder packing with spheres that are in contact with each other but that have no mutual intersection (Fig 2.26(a)). It would make no sense to use fewer spheres than what corresponds to $\lambda=2$, as the resulting packing would be totally unbalanced (Fig 2.26(b)). We have taken $\lambda=1$ by default since it gives an interior representation with acceptable density.

If m is the number of circles that make up the generalized packing for the generating polygon of the solid, the total number of spheres in the interior representation will be given by

$$\Sigma n_i,$$

where the sum extends for $i=1, 2,..., m$.

The previous discussions are no longer valid when we have an object whose h value is less than the diameter of the first element C_1 of the set C_∞. This circle has the greatest radius of all the circles in the packing for polygon B; indeed, we know that there cannot exist a circle inside B that is greater than the circle with the greatest radius of the set of maximal circles C^m, and that — because of the construction scheme— this circle is the first element of C_∞. Then, as the sphere S_1 that defines the set of translates corresponding to C_1, has a radius r_1 equal to the radius of C_1, and since $2r_1>h$, it becomes evident that it is impossible that S_1 is part of a valid packing for the solid in question.

Similarly, if

$$2r_1 \geq 2r_2 \geq ... \geq 2r_k > h,$$

the k spheres $S_1, S_2,..., S_k$, would not be well defined as they are not contained within the represented object.

Therefore, a specific approach for this type of *flat* object is required. The proposed solution consists in not introducing a unique cylinder for each circle C_i with $2r_i > h$; rather we will define several cylinders instead, so that their corresponding spheres S_{ij} will have radii r_{ij} such that $2r_{ij} = h$. To do this, it will be necessary to define an algorithm that allows us to obtain a generalized packing of a circle by means of circles.

Now, our problem can be stated as the following: given a circle C whose radius has any value R, we want to define a set of circles

$$c_1, c_2,..., c_n,$$

all of them with a radius equal to a prefixed value r, in such a way that they define a packing for C.

First, (Fig. 2.27) we build a ring formed by n_1 circles whose centers are uniformly distributed on a circumference concentric with C and with radius

$$R_1 = R - r,$$

so that

$$n_1 = excess\ (2\pi R\ /\ \lambda_c\ r).$$

Next, we define a circle C with radius

$$R_2 = R - 2r,$$

so that if $R_2 > r$ a second ring will be defined by merely substituting R_2 for R. We do this successively until either $R_t < 0$ —which would mark the end of the process— or $R_t \leq r$. In this last case, adding another circle with radius r and concentric with C would suffice.

The coefficient λ_c has an important role analogous to that of λ, that is, it serves to fix density, in this case for the local packing of a circle. It may vary between the same limits $(0 < \lambda_c \leq 2)$, in such a way that the number of circles in the ring increases as λ_c decreases. If

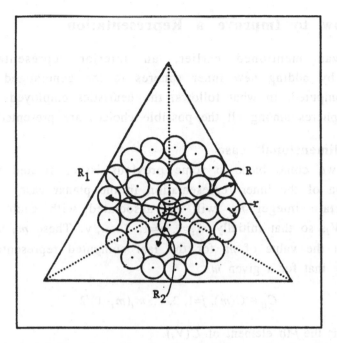

Figure 2.27   A packing for the special case of a flat object

$\lambda_c=1$, two contiguous circles will be placed in such a way that the distance between their centers would be approximately equal to r; for $\lambda_c=2$ this distance would be of the order of $2r$; and if λ_c tends to zero we would have infinite spheres with almost coinciding centers.

An additional heuristic criterion is based on the fact that, according to the construction scheme, each circle of C_∞ is tangent to at least two polygon edges, and then —for the cases in which n_1 is a low number— two or three circles have been defined with radius r situated in such a way that they are tangent to the same edges as the original circumference, thus guaranteeing no quality loss in the packing.

With this, the description of the interior representation for the three-dimensional case comes to a close. We will now move on to consider the quality question of an interior representation and the spheration process.

2.2.3 How to Improve a Representation

As was mentioned earlier, an interior representation is improved by adding new inner spheres to the generalized packing already computed. In what follows, the heuristics employed to select the best spheres among all the possible choices are presented.

a) Two-dimensional case

Now we come back to some previous issues related with the construction of the inner representation in the planar case.

A certain integer m_i will be associated with each Voronoi segment V_i, so that initially $m_i=1$, $i=1, 2, ..., v$. These m_i will be a function of the value of m for the last computed representation, in such a way that for a given m:

$$C_{ij} \in C(m), j=1, 2, ..., m_i(m_i-1)/2,$$

where C_{ij} is the j-th element of $C(V_i)$.

If V_i is a terminal segment, another parameter d_i is also associated with it (see Fig. 2.28) as defined by:

$$d_i = \min [d(A_i, D_{ij}) - r_{ij} ; j=1, 2, ..., m_i-1]$$

where A_i is the end of V_i that is also a vertex of the polygon, r_{ij} the radius of C_{ij}, and D_{ij} its center.

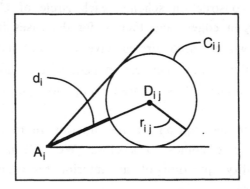

Figure 2.28   Parameter d_i associated with a terminal segment

Then, the worst Voronoi segment is chosen according to the following steps:

1.- The terminal segment with the largest value for d_i/m_i is selected;

$$\delta = \max\ [d_i/m_i\ ;\ i=1, 2, ..., p]$$

where p is the number of terminal segments. Let V_t be the selected segment, worst terminal segment (for $i=t$)

2.- The non-terminal segment with the largest value for l_i/m_i is selected;

$$\varepsilon = \max\ [l_i/m_i\ ;\ i=p+1, p+2, ..., v-p]$$

where $l_i=d(A_i,B_i)$ is the length of V_i. Let V_s be the selected segment, worst non-terminal segment (for $i=s$).

3.- Finally, if

$$l_t/n_t \geq \kappa\varepsilon;$$

then V_t is the worst Voronoi segment; else V_s is the worst. κ is a coefficient ($\kappa<1$) that favors the choice of the terminal segment rather than the non-terminal, due to reasons that will later become apparent.

Once the worst V_i has been selected, m_i is incremented by one and $k=m_i-1$ circles from $C(V_i)$ are added to $C(m)$. Its centers are distributed in a certain way along V_i between A_i and B_i. To define this distribution, two cases must be distinguished:

a) If the segment is V_s, non-terminal, the centers are homogeneously scattered along the segment, that is, they will be a distance $\varepsilon=l_s/n_s$ apart.

b) If the segment is V_t, terminal, the centers will be located at successive distances given by

$$\delta l = l_t \log\ (1 + 9k/m_i);\ k=1, 2, ..., m_i-1,$$

these distances are measured from B_t, the end of the segment that is a Voronoi point (Fig. 2.29).

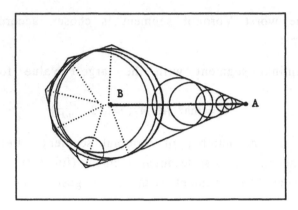

Figure 2.29   Distribution of centers on a terminal segment

To justify our latter statements we have to turn back to the heuristic criteria that have motivated them. The distinction between terminal and non-terminal segments is a consequence of a frequently observed fact: the number of circles that must be placed on a non-terminal segment to get a good representation is considerably smaller that this number for a terminal segment with the same length. Two reasons account for this fact:

(1) Terminal segments are always in the interior of the polygon, far from its vertices and edges.

(2) As a consequence, circles centered on them will have comparatively larger radii.

Then, few of these circles will suffice to fill an important portion of the polygon. Factor κ, serves to favor the selection of terminal segments as the worst. Moreover, the criteria for the selection of worst terminal segment and the worst non-terminal one have been different: d_i/m_i has been used for the former and l_i/m_i for the latter.

This last fact is related with another heuristic criterion that has been used to promote those circles located at the proximity of the vertices of the polygon (Fig. 2.29), since in these zones circles will necessarily be smaller and will fill a smaller portion of the polygon interior. That is why a uniform distribution has been employed for circles on non-terminal segments, while a logarithmic distribution has been applied, with good results, for the terminal case.

b) Three-dimensional case

For reasons that will only become apparent when application considerations are discussed, a criterion has been selected that favors those packings that are evenly spread over the whole volume of the solid. Moreover, special attention has been paid to guarantee that the regions in the surroundings of the vertices and edges of the object are properly represented.

At first, we could have followed a scheme similar to the one developed for the exterior case, making a partition of the solid in different areas and assigning a partial packing to each one, so that later, by means of the expert spherizer, we can decide which of the different possible actions are the most appropriate. Nevertheless, a less complex method has been followed in hopes of simplifying the representation. This decision will not be totally justified until the next chapter where we consider the application of the representation. Then it will be seen that, in some way, the interior representation has a secondary role subordinate to the exterior representation. This fact makes such a detailed description of it unnecessary. More concretely, for the collision detection problem, the exterior spheres serve to guarantee that no collisions exist, whereas the interior ones are used to ensure that they do. Naturally, to keep objects from crashing into one another, it will be absolutely necessary to ensure that there are no collisions, while knowing with certainty that there is a collision will help to accelerate the decision-making process.

Accordingly, first an adequate set $C(m)$ is computed. The systems of translates

$$\{S_i + \mathbf{a}_{ij}, j=1, 2, ..., n_i, i=1, 2, ..., m\},$$

will be generated from it; they make the complete packing for the object. For a given value of i, system $\{S_i + \mathbf{a}_{ij}\}$ will be composed of n_i spheres. As we have already seen (in section 2.2.2), n_i depends on h, r_i (radius of S_i) and λ. As h does not change, if λ is kept constant, the number of spheres for the system generated by S_i also remains invariant.

Consequently, to improve a given interior 3D representation, we refine the generalized packing for the generator polygon of the solid, according to the schemes outlined in the precedent section: making $C(m)$ larger by advancing in the sequence C_∞. For each new S_i the spheres in its system $\{S_i + \mathbf{a}_{ij}\}$ are generated. The density coefficient λ has been kept invariant (and with a default value $\lambda = 1$ in our implementation) for all the representations of an object, it is only changed in case we wish to *tune* the system.

2.3 Convergence and Hierarchy

The first aim of this section is to demonstrate that the double representation that has been presented has the important property of convergence. This concept has to be understood in the framework of the sequential scheme that allows us to define a succession of representations in such a way that each one improves upon its predecessor. Our result is that a representation with an error as small as desired can be obtained by taking subsequent elements in the sequence of representations. As a sequel of this property a second goal will be attained: a double hierarchy of representations can be defined to describe the solid object at different levels of accuracy.

We must insist on the importance of the convergence of our representation, as this is the distinctive nature of our model that makes it different from any other ways of handling the representation of solids by means of spheres. If this were not one of its properties, it could be argued that the spherical representation is interesting but it is doomed to be crude, limited by the ease in which the shape of a certain object lends itself to *conform* to a set of spheres. Nevertheless — with the limitations imposed on the represented solids— our spherical model is totally general, in a way that the object may be represented as exactly as is desired by the sole use of spheres. Naturally, some objects will adapt to the model better and will need fewer spheres than others to reach the same precision.

2.3.1 Convergence of the Exterior Representation

Let us now examine the convergence issue for the case of exterior representations. First some results for particular examples are examined in order to come to general conclusions later. Figure 2.30 shows a sample listing of the expert output for the first 11 representations of the column-like object shown in Fig. 4.11.

```
TOTDEL SIDEWE SIDEP TOPDEL TIPWEI TIPEP FALSID FLTIP NTOT NSI NTO NTI CS CT
================================================================================

 1 Initial STARTUP     No rule applied, just after initialization

8.437  9.492  1.16   .000   .000   .00   .751   .00    1   1   0   0   1   0

 2 (4.1)   MORESPHERES Improve side: Correct delta

3.060  3.242  1.16   .000   .000   .00  1.363   .00    2   2   0   0   1   0

 3 (4.1)   MORESPHERES Improve side: Correct delta

2.133  2.084  1.16   .000   .000   .00  1.832   .00    3   3   0   0   1   0

 4 (4.1)   MORESPHERES Improve side: Correct delta

1.849  1.679  1.16   .000   .000   .00  2.177   .00    4   4   0   0   1   0

 6 (2.2.1) MAKETIP     Correct side: there is no top, but side is worst

1.675  1.288  1.16   .000  1.679  1.16   .000  2.18    6   4   0   1   1   1

 7 (1.2)   MORECIRCLES Correct tip: there is no top, but tip is worst

1.441  1.288  1.16   .000  1.466   .80   .000  1.49    8   4   0   2   1   2

 8 (1.2)   MORECIRCLES Correct tip: there is no top, but tip is worst

1.478  1.288  1.16   .000  1.767   .98   .000  1.12   10   4   0   3   1   3

 9 (0.2)   MAKETOP     Top is needed, tip already exists

2.024  1.288  1.16  4.167  1.767   .98   .000  1.12   12   4   1   3   1   3

10 (5)     MORECIRCLES Improve top, top is the worst

1.880  1.288  1.16  3.276  1.767   .98   .000  1.12   14   4   2   3   1   3

11 (5)     MORECIRCLES Improve top, top is the worst

1.797  1.288  1.16  2.768  1.767   .98   .000  1.12   16   4   3   3   1   3
```

Figure 2.30   Sample output listing of expert spherizer

Each step in the spheration process produces two data lines. The first line contains three items describing the applied action:

    a) Representation number or step in the process.
    b) Code identifying the applied rule (in brackets).
    c) Brief description of action and purpose of the rule.

The second line includes some data to characterize each representation:

$$\delta_{tot}, \; \delta_s + \varepsilon_s, \; \varepsilon_s, \; \delta_{to}, \; \delta_{ti} + \varepsilon_{ti}, \; \varepsilon_{ti}, \; \delta_{fs}, \; \delta_{ft}, \; N, \; N_s, \; N_{to}, \; N_{ti}, \; m_s, \; m_{ti}$$

The most important data for this example have been plotted in Fig. 2.31: namely, δ_{tot} coefficient (*total* δ) versus the sequence of representations and δ_{tot} versus the total number of outer spheres (N) in each representation (δ_{tot} is expressed as a distance in dm). A representation is identified by its ordinal number in the series; naturally, there exists a correlation between this number and the number of spheres composing the representation.

It can be seen that the value of *total* δ tends to be smaller and smaller as we advance along the sequence of representations. Moreover, convergence is very fast at the beginning, a steep initial slope can be seen with a local minimum for representation number 7 with $\delta_{tot} = 1.441$ dm. If we define a relative error as the ratio of the average error distance δ_{tot} with respect to a characteristic size for the object (the diameter of the smallest enclosing sphere for the whole object, 28.17 dm in this case), then we obtain an error of 5.11%. It has to be pointed out that representation number 7 uses only 8 spheres. The best shown result is for representation number 100 with a 3.07% relative error. Naturally, convergence slows down as we advance further in the sequence, but always tending towards smaller values.

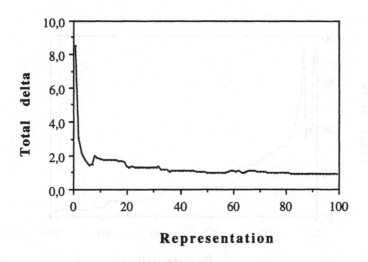

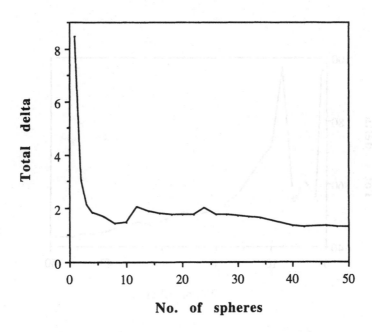

Figure 2.31   An example of the evolution of δ_{tot} quality coefficient

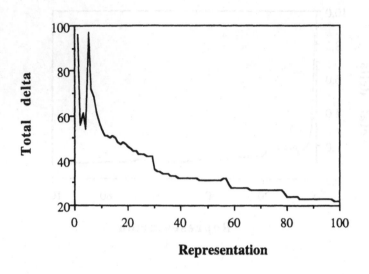

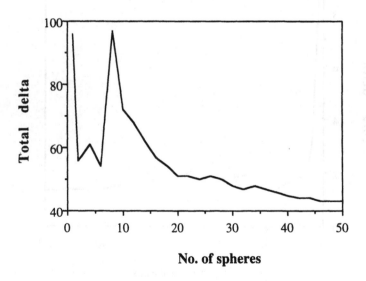

Figure 2.32 Evolution of δ_{tot} for a particularly bad case

The ups and downs in the plot can be explained considering the way the spherizer operates. First, some actions add many spheres to the new representation, while some others slightly increase the total number of spheres (adding just two). Obviously the former will change the quality more than the latter. In addition, the latter actions may temporarily worsen the representation because a given representation may not be made better in just one step. Consider, for instance, the case of a regular 12-sided polygon being represented with two circles; that cannot obviously mean an improvement over the one-circle representation. But in any case the quality always becomes better after further applying some more rules, often the same one.

The previous situation is also instanced in Fig. 2.32 that shows a particularly bad case (for object shown in Figs. 2.3 and 2.16). Here after a first minimum for representation number 4 (9.81% relative error with 6 spheres), there is a high peak. This is due to the need for creating the top covering. The first covering for the top must be inevitably of poor quality as it uses only one sphere. Note that this action is necessary to guarantee the correctness of the representation, even though it may make it worse. In general, better results are obtained for elongated objects than for flat ones.

From this example and other similar results for various test cases, we can conclude that the sequence of representations obtained as output from the repetitive application of the rules of the Expert Spherizer converges towards a perfect no-error representation. That is, if a sequence of real number is defined as composed of the values of δ_{tot} for the successive representations

$$\{\delta_{totn}\} = \{\delta_{tot1}, \delta_{tot2}, \delta_{tot3}, \delta_{tot4}, \delta_{tot5}, ...\},$$

then, this is an infinitesimal sequence, that is, it is convergent towards a limit equal to zero:

$$\lim_{n \to \infty} \delta_{totn} = 0$$

Sequence $\{\delta_{totn}\}$ is, however, not monotonic since it does not stand that

$$\delta_{toti} \geq \delta_{toti+1}, \ \forall \ i=1, 2, ...$$

This fact is of no importance since it is clearly a Cauchy sequence. Indeed, for every real ε as small as desired, there exists a certain representation n_o such that

$$|\delta_{toti} - \delta_{totj}| < \varepsilon, \quad \forall\ i \geq n_o, j \geq n_o.$$

That is, as we move in the sequence of representations, the values of δ_{tot} come closer and closer to one another.

The result of our experiments with different types of objects is that an error of less than 10% can be generally obtained with an outer representation composed of less than 20 spheres. This representation corresponds to the first minimum of δ quality coefficient. Therefore, most of the time an object will be represented by means of a set of less than 20 spheres. Only in special situations more spheres will be needed, even hundreds if necessary, since the model does not limit the maximum number of spheres in the representation of an object.

Of course, although a non-monotonic sequence has been presented to show the construction method, the representations that do not improve on their predecessors are discarded for practical applications, and only a decreasing monotonic subsequence of representations will be used.

2.3.2 Convergence of the Interior Representation

A similar result can be established for interior spheres. In the planar case convergence is evident as it is an immediate corollary of our result concerning the generation of a perfect unity-density packing in 2D.

An error set E can be defined to characterize an inner representation as in the exterior case:

$Q \in E$ if and only if $\exists\ O_i \in O\ /\ Q \in O_i$ and $Q \notin S_j, \forall j\ (j=1, 2, ...)$,

or simply, $Q \in O$ and $Q \notin S$, or just, $E = O - S$.

Where:
$O = \{O_i,\ i=1,...,m\ \}$ is the set of all objects and
$S = \{S_j,\ j=1,...,n\ \}$ is the set of all spheres in the representation.

In the planar case convergence is evident as we have already shown the existence of a perfect representation:

$$C_\infty = \lim_{m \to \infty} C(m).$$

If we denote by $E[C(m)]$ the error set associated with the mth representation, it is evident that

$$\lim_{m \to \infty} E[C(m)] = E[\lim_{m \to \infty} C(m)] = E[C_\infty] = \varnothing;$$

resulting in an infinitesimal sequence. In addition, it is monotonic:

$$E[C(m)] > E[C(m+1)] \ \forall \ m=1, 2, \ldots,$$

since adding more spheres is a direct way of approaching the perfect, final packing.

Similarly, the extension to 3D can be shown to be also convergent, but in its present definition it is not infinitesimal. To be so, a special treatment for the tips will be needed, in the fashion of the outer case, but that would render an unnecessarily elaborate model.

Chapter 3

Collision Detection

The first application of the spherical model is the problem of collision detection, that is, how to recognize a motion that is safe. The solution to this problem is a precondition to deal with the problem of motion planning, but it can also be regarded as a problem in its own right. First, *off-line* robot programming systems are based on interactive computer simulation systems: a system which detects collisions during this simulation process is required as a tool to facilitate the programmer's job who, otherwise, would have to visually detect collisions. Secondly, it is evident that in order to plan the movement of a robot so as to avoid collisions, a method which detects when such a collision takes place should necessarily be available.

We will begin this chapter by posing the different ways in which the problem can be stated and a summary of the main solution methods. Then, we will present our solution model based in the twofold spherical representation; to do so, we will deal first with the static case, then we will consider the dynamic case, and finally we will analyze the particular problem of detecting collisions for a robot manipulator in motion.

3.1   Problem   Definition

The question we are about to take on is a part of a wide-ranged set of problems that can be grouped as *proximity and intersection problems*. In an initial classification, we can divide these problems into two large groups, according to whether or not they admit the possibility that some of the objects involved in the problem move. Then, one speaks about the static detection of intersections on the one hand, and dynamic detection on the other, as well as the static and dynamic proximity problems. Even though the terms intersection and collision are sometimes used interchangeably in this context, throughout this book, we will refer to the static problem simply as *intersection detection* and we will restrict the use of the term *collision detection* to the dynamic problem.

3.1.1   Problem   Classification

Let us, then, have a look at the different problems which subdivide the two main categories mentioned.

Intersection Detection. All involved objects are at rest. As was done in the spherical representation study, to define an object, we will consider it as a set of points in space, therefore, we will know the exact position of each and every one of the points that make up the object at a given moment; and in this particular case, being a static problem, these positions will not depend on time.

(1) The simplest case is the one that only involves two objects B and C, in such a way that we can define the problem by saying that it deals with determining if their intersection is or is not the empty set

$$B \cap C = \emptyset.$$

A problem that goes beyond mere intersection detection is the computation of the object obtained as the result of the intersection, when this intersection is not empty (*intersection computing* problem)

(2) A second problem is the one where we have a certain set

$$S = \{ S_i, i = 1, 2, ..., n \}$$

that is made up of n objects, and we want to know whether there is intersection between any two elements of this set, that is, if it is true that

$$\forall S_i, S_j \in S, \ S_i \cap S_j = \emptyset;$$

if we also need to know the number of these intersections (the *counting intersections* problem), the complexity increases and if we want to report which are the pairs of objects (S_i, S_j) whose intersections are not empty we make it even more complex (the *reporting intersections* problem)

(3) The problem becomes more complicated if several sets of objects appear and we assign the same *color* to all of the objects that belong to the same set, so that two different sets have two different colors. Then, we are dealing with determining whether there exists some pair of differently colored objects that intersect one another, or rather

$$S = \{ S^j, j = 1, 2, ..., n \},$$

where in turn

$$S^j = \{ S^j_i, i = 1, 2, ..., n_j \},$$

and we want to find out if

$$\exists S^j_i, S^k_l \ ; \ S^j_i \in S^j, \ S^k_l \in S^k \ (j \neq k) \ | \ S^j_i \cap S^k_l \neq \emptyset;$$

(4) Another question that we can pose in the previous cases —if there is no intersection— is how to solve the so-called *proximity* problem. In the simplest case, with only two objects B and C, this would consist in computing the distance between them, defined as the smallest distance between any two points of each object:

$$d\ (B, C) = inf\ \{\ d\ (M, N);\ M \in B,\ N \in C\ \},$$

that reminds us very much of the definition of *margin*, that will be stated in chapter 4, with the mere substitution of one of the objects for a certain placement of the robot.

(5) In the case of having a set of objects with no mutual intersection, similarly we can ask ourselves which is the minimum distance between all pairs of objects

$$inf\ \{\ d\ (S_i, S_j);\ S_i, S_j \in S,\ j{\neq}i\ \};$$

another question that is also of interest is to determine for each object $S_i \in S$, which is the object of S that is closer to S_j.

(6) Finally, for several sets of objects, the proximity problem can be stated by saying that we want to know the least distance between differently colored objects, naturally supposing that there exists no intersection between any of them. This can be stated, using the previous terminology, as

$$inf\ \{\ d\ (Sj_i, Sk_l);\ Sj_i \in Sj,\ Sk_l \in Sk,\ j{\neq}k\ \};$$

Collision Detection. Now we accept that some of the objects involved in the problem can vary their position. That is, for each object B —defined by a set of points— we will suppose that there exists a *placement* function $F(t)$ so that for a particular instant in time t_i, the function $F(t_i)$ represents a transformation of object B from its initial position to the position that it occupies in time t_i, in such a way that the new resulting set $F(t_i)B$ will be made up of points defined as

$$F(t_i)B = \{\ Q;\ \exists\ P \in B \mid Q = F(t_i)P\ \}.$$

Similarly to the static case, we can point out different problems that grow in complexity as the situation they deal with gets more and more complicated.

(7) The simplest case is the one that considers two movable objects B and C, and we want to establish if for a particular path

—given, for example, over a time interval $[t_0, t_f]$— there is any collision between the two bodies, that is, if there exists some instant $t_i \in [t_0, t_f]$ for which it is true that

$$F(t_i)B \cap F(t_i)C \neq \varnothing,$$

appearing an intersection between B and C in that moment.

(8) In a similar manner we would have the problem of determining if there is a collision between objects belonging to a particular set S during a particular interval of given time. In this case, we must guarantee that

$$\forall \ S_k, S_j \in S \ \text{ and } \forall \ t \in [t_0, t_f], \ F(t)S_k \cap F(t)S_j = \varnothing;$$

so that no collision occurs whatsoever.

(9) We could also consider various sets of objects —each one with a different color— and find out if two differently colored objects collide when they follow their paths in a given period of time, so that if

$$\forall \ S^{j}{}_i \in S^j, \ S^{k}{}_l \in S^k \ (j \neq k) \text{ and } \forall \ t \in [t_0, t_f], \ F(t)S^j{}_i \cap F(t)S^k{}_l = \varnothing;$$

is fulfilled; thus we can ensure that no collision occurs

(10) Finally, for the last three cases —if there is no collision— we can consider the proximity problem. To do this, we will define the distance between two solids B and C during a certain movement μ —as given by $F(t)$— for the interval $[t_0, t_f]$ as

$$d\,(B, C, \mu) = inf \ \{ \ d\,(F(t)B, F(t)C); \ t \in [t_0, t_f] \},$$

that is similar to the definition of the margin of a motion.

(11) Similarly, for the case of set S, the minimum distance between all involved objects in motion, will be given by

$$inf \ \{ \ d\,(F(t)S_i, F(t)S_j); \ S_i, S_j \ \in S, \ j \neq i; \ t \in [t_0, t_f] \}.$$

(12) Finally, for various *colored* sets we may be interested in knowing the value of

$$inf \{ d \ (F(t)Sj_i, F(t)Sk_l); \ Sj_i \in Sj, \ Sk_l \in Sk, j \neq k; \ t \in [t_0, t_f] \}.$$

that represents the minimum distance between two differently colored objects throughout the entire movement considered.

3.1.2 Solution Methods

To analyze the different techniques that have been proposed for offering solutions to the problems we have just mentioned, we must keep in mind that these techniques are highly conditioned by the method used to represent the objects involved in the problem. Next, we will briefly enumerate the advantages and disadvantages of the most usual models for representing solids.

Constructive Solid Geometry (CSG). Even though CSG has not classically been used for collision and intersection detection, some systems have employed it successfully ([Cameron, 1989], [Faverjon, 1989]) since by their very definition, they give rise to a hierarchical representation that easily lends itself to a division into simpler subproblems with the purpose of reducing the complexity of the total computation.

Boundary Representation (B-rep). This is one of the first models used for the static as well as the dynamic problem [Boyse, 1979]. It has also been frequently used later on ([Canny, 1986], [Kawabe, Okano and Shimada, 1988], [Bobrow, 1989]) due to the fact that intersection conditions can be easily formulated from relationships between edges and faces. Some of the most well-known systems for motion planning are based on this model for collision detection [Lozano-Pérez, 1987], [Donald, 1987]. In general, its use will give rise to not very efficient algorithms for real designs in which polyhedra with a large number of vertices will habitually become involved.

Cell Decomposition. In theory this technique may be considered as an equivalent to that of boundary representation from the point of view of its adaptation to the solution to the intersection problem. As no superior hierarchical structure exists, since the decomposition is unique, cells will have a role equivalent to edges and faces of the boundary representation.

Spatial Occupancy Enumeration. This is also widely used, especially in the realm of computer graphics, in order to solve the *interference* problem —the name by which the intersection problem in this realm is known— since, by use of *voxels* —cubes with fixed size and position— it is easy to determine mutual intersections. This method tends to be used for the static case and it only handles collisions by means of methods which reduce to intersection computations (the multiple intersection detection method). Its main disadvantage is that it generates a large storage volume and it gives rise to non-efficient algorithms.

Octrees. This can be considered as a hierarchical version of the spatial enumeration technique, consequently improving the behavior of the latter as it does not need so much storage space and it is more efficient. The octree method is basically one in which a cubic element is divided into eight smaller cubes so that each one of them is subdivided again giving rise to a tree structure with a branching factor of eight. Each node is labeled according to its relative position with respect to the represented solid: full, empty, or mixed; and only mixed octrees will be recursively subdivided. Even though it is adequate for the intersection problem [Moore and Wilhelms, 1988] in the collision case this method has important disadvantages [Hayward, 1986], [Dupont, 1988] especially due to the fact that transformations implicit to movement — translations and rotations— are computationally costly as they mean recalculating all the nodes of the representation tree.

Swept Representation. With respect to the more enumerative models already cited, this one has the disadvantage that the necessary techniques of analytic geometry involved may be excessively complicated thus making it difficult to determine the intersection between two solids.

For a systematic study of the different solution methods for the intersection and collision detection problem, we should make a logical distinction between the static and dynamic case. In the first case, nevertheless, it is difficult to propose a clear classification of the existing algorithms, therefore —keeping in mind that our interest focuses on robots *in motion*— we will limit ourselves to present now three fundamental methods in which techniques dealing with the collision detection problem can be classified, (problems 7-9).

a) Multiple Intersection Test

This method consists in reducing the collision detection problem during a motion to solving a sequence of intersection detection problems; and it is sustained by the fact that the static problem is obviously less complicated than the dynamic problem. In essence, we have to verify the absence of geometric intersections for particular instants $\{t_i\} \in [t_0, t_f]$ of the total time interval, thus transforming a continuous problem into a discrete problem. In this way, we could detect a collision like the one shown in figure 3.1 in which each image corresponds to an instant t_i in those where intersection has been verified.

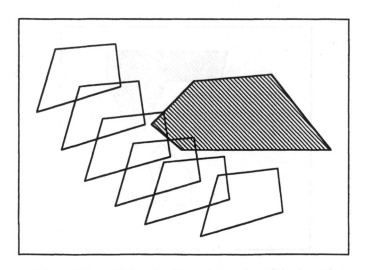

Figure 3.1   This collision is detected by a multiple intersection test

It is evident that to have some degree of validity, this method must guarantee that the time interval Δt_i between two consecutive sample instants

$$t_{i+1} = t_i + \Delta t_i$$

is small enough so that there is no collision between positions in t_{i+1} and t_i that would go undetected, as shown in Fig. 3.2. On the other hand, if this Δt_i is too small, the algorithm suffers from an excess of computation time.

An initial criterion to fix Δt_i is to divide the total time interval into equal parts

$$\Delta t = (t_f - t_0)/n,$$

where n is *adequately* fixed [Meyer, 1981]. In general this option is the most widely used in computer graphics applications in which the criterion to fix n is habitually conditioned by the speed of the animation expressed in images per second [Uchiki, Ohashi and Tokoro, 1983], in such a way that Δt would be the difference between the times corresponding to two successive images. Then this

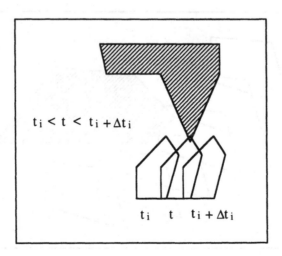

Figure 3.2   This collision, however, is not detected by a multiple intersection test. Δt is not small enough

Δt must be such that the fastest object does not travel in time Δt a distance greater than the resolution of the system (a *pixel* for visualization effects). Note that in these cases to solve the dynamic problem we will have to solve a number of intersection problems of the order of several hundreds or even thousands.

Other more serious alternatives for the determination of Δt_i have been considered by [Cameron, 1985] starting from a computation of the minimum distance between any two objects and an estimation of the speed of the objects, in such a way that it uses such quantities as

$$\text{distance} \,/\, \text{time}$$

to guide us in obtaining Δt_i. Note that this implies solving not only the intersection problem, but also the proximity problem as well —in the previously considered cases (4-6)— for each instant t_i, thereby notably complicating the problem and increasing computation time.

b) Swept Volume Intersection Test

This method is based on generating the volume swept by a three-dimensional solid during its motion. To each object B involved we associate a new three-dimensional solid SVB defined by its swept volume

$$SVB = \{\ F(t)B \ ; \ t \in [t_0,\ t_f]\ \};$$

that is, it is defined by the set of points in space that have belonged to the solid in some moment of the time interval considered. Then, the absence of collision between two objects in motion, B and C, is a sufficient condition to know that no intersection exists between their corresponding swept volumes

$$SVB \cap SVC = \varnothing.$$

We see that once again we are dealing with transforming the dynamic problem into a static problem, with the particular feature that in this case we would detect collision by means of the solution to a single intersection problem. Therefore, the dynamic problems (7), (8) and (9) would respectively turn into the static problems (1), (2) and (3) by merely substituting in the corresponding equations that define them $F(t)B$ with $t \in [t_0, t_f]$ for SVB, for each object B involved.

This simplification of the problem, naturally, is not achieved without introducing some additional complications. First, it is generally difficult to obtain an explicit representation of the swept volume. Secondly, the existence of intersection between the swept volumes of two objects is not a sufficient condition for collision between them, since it is evident that the object could have occupied the same space at different instants. Moreover, the necessary temporal information has been lost, since it does not appear in the generation process of the swept volumes.

To solve the first problem —the difficulty for explicitly obtaining the swept volume— two solutions have been proposed:

1.- To restrict the set of solids involved, as well as their motions so that we can easily represent the swept volume. Such is the case of the LARS system [de Pennington, Bloor and Balila, 1983] which handles spheres that can move following rectilinear paths or arcs of circumference, with which they generate cylinders and toruses, respectively.

2.- To work only with an implicit representation of the swept volume; in this way [Boyse, 1979] presents a method which detects collision between polyhedra generated by sweeping each element —vertices, edges and faces— individually to study later on the possible intersections between the sweeping of those elements each one separately, without trying to construct an explicit joint representation of the total swept volume at all.

To overcome the second disadvantage —the necessity of a condition that suffices for collision— we have two alternatives:

1.- To consider that the total motion can decompose in such a way that of all possible sets of differently colored objects (problem 9), there are no two that are simultaneously in motion, so that the collision will always occur between an object in motion and another at rest, and the sufficient condition would be guaranteed.

2.- The other possibility is to consider the relative motions of each
   pair of differently colored objects, which is not always easy;
   moreover, if there are many objects in motion, it becomes
   necessary to take many such relative motions into account.

c) Four-Dimensional Intersection Test

This method solves the problem from a different point of view:
instead of working in ordinary three-dimensional space, one more
dimension is added —namely, time— to reduce collision detection to
a static problem in a space with four dimensions: three spatial
coordinates and time.

In this way, for each real object we define a set of points in that
four-dimensional space that serve to describe the different positions
that the points of the object have occupied for each instant in time t
$\in [t_0, t_f]$. To understand this idea we can represent a two dimensional
example (Fig. 3.3a) in which we want to detect collisions between
two polygons moving on a plane. Then, we can imagine this plane as

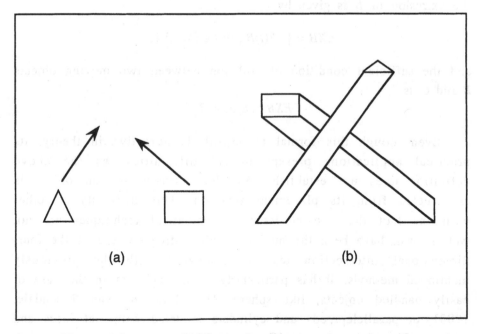

(a)   (b)

Figure 3.3   Example of extrusion for two polygons in motion

if it were the floor of an elevator that rises with constant velocity, in such a way that, combining the motion of the polygon on the floor and the upward motion of the elevator the polygons will generate some three-dimensional figures by sweeping (Fig. 3.3b). If we consider the vertical dimension as time, a section by a horizontal plane would give us the situation of the polygon for a particular instant.

It is evident that a necessary and sufficient condition for the existence of collision between two objects is that the intersection between their corresponding four-dimensional set (or the three-dimensional ones in the example in the figure) is not null. Indeed, if two points coincide in four dimensional space, their coordinates (x, y, z, t) will be the same, which means that the original objects have shared a particular portion of space at the *same* time, in other words, they have collided.

Formally, the operation that allows us to obtain the four-dimensional set from a solid B and its motion as given by the placement function $F(t)$, is known as *extrusion*, in such a way that the extrusion of B is given by

$$EXB = \{ (F(t)B, t); t \in [t_0, t_f] \};$$

and the sufficient condition of collision between two moving objects B and C is

$$EXB \cap EXC \neq \emptyset.$$

Even though this formal framework is attractive in theory, its practical applications present logical difficulties, as the above definition does not establish how the extrusion of an object is constructed from its placement function. This difficulty, together with the fact that the necessary mathematical techniques are not well known, have been the motive for the infrequent use of the four-dimensional intersection test as compared with the previously mentioned methods. It has particularly been employed in the case of easily handled objects, like spheres [Esterling and van Rosendale, 1983], or parallelepipeds and cylinders modeled with flat faces and with rectilinear movements [Cameron, 1990].

3.2 Intersection Detection

As a first application we will now present a technique based on our spherical representation to solve the intersection detection problem. Before continuing, it is necessary to remember that we are dealing with a representation which has the following properties:

(i) The only elements involved in the representation are spheres, so our problem will always be reduced to handling sets of spheres, whatever the real object underlying each spherical representation may be.

(ii) We have two sets of spheres, so that one of them —the exterior representation— is made up of spheres that cover the boundary of the represented object. The other set — the interior representation— is made up of spheres that are found entirely within the interior of the object.

(iii) For each one of these two representations we have a hierarchy of sets which make the approximation of the object get better and better —its exterior boundary or its interior part— using more and more spheres. For each representation we have some parameters which, globally or locally, give us a measure of representation accuracy.

Depending on which of these properties we take into account, we will be able to define different algorithms. That is, if we only take property (i) into account, we can propose algorithms that handle sets of spheres but that do not distinguish whether they are exterior or interior; and if we only ignore property (iii), we will compare fixed sets of interior and exterior spheres but we will be unable to consider the possibility of resorting to other better representations that solve indeterminate cases.

Next, we will consider different types of problems and offer algorithms that solve each case. We will make use of our hierarchical spherical representation to a greater or lesser extent depending on the cases; that is, we will use all or some of the above mentioned properties.

3.2.1   Intersection   Between   Spheres

The problem of determining the existence of intersection between a group of spheres by comparing all possible pairs of spheres is an immediate method that has been sometimes used. The reason for this is that in many of the approaches already mentioned, there is an attempt to improve the efficiency of the proposed methods by making a preliminary approximation using boxes (rectangular parallelepipeds) or enclosing spheres that completely enclose an object. This is done since, by their very nature, the direct application of those methods means doing a large number of operations. Therefore, this simplification is proposed, for example, to alleviate the computational load of techniques based on boundary representations [Boyse, 1979], spatial occupancy enumeration [García-Alonso, Serrano and Flaquer, 1994], or cell decomposition [Moore and Wilhelms, 1988]. In all of them, to keep us from having a large number of comparisons between vertices, faces, edges, voxels or cells, we first compare the containers enclosing each one of the objects in order to be able to quickly identify easily detectable non-intersecting cases.

a) Comparison Between Boxes and Enclosing Spheres

Sometimes, when choosing between enclosing boxes or spheres, arguments have been put forward which try to show the supremacy of the first mentioned over the latter. These arguments, nevertheless, are fallacious for reasons which we will state in the following. Furthermore, we will show that in equal conditions, spheres are always more appropriate containers than boxes. In essence, the confusion arises from associating the application of enclosing boxes with the minimax test which is usually used as a quick way of determining the absence of intersection between boxes. In the following paragraphs, we will suppose that a quadratic algorithm will be used and therefore, the determining factor for evaluating complexity will be the number of operations necessary for detecting the intersection between two objects.

The minimax test gives six sufficient —but not necessary— conditions for non-intersection between two objects. This consists in determining the maximum and minimum values of the three coordinates for all the points that make up an object, so that we associate to each object six *minimax parameters* $(x_m, x_M, y_m, y_M, z_m, z_M)$ that define the smallest parallelepiped with parallel sides to the coordinate axes and that totally encloses the object. Then, the six sufficient conditions for non-intersection between two objects B and C are

$$x_{BM} < x_{Cm}$$
$$x_{CM} < x_{Bm}$$
$$y_{BM} < y_{Cm}$$
$$y_{CM} < y_{Bm}$$
$$z_{BM} < z_{Cm}$$
$$z_{CM} < z_{Bm}.$$

Obviously, having one of these conditions satisfied is sufficient to be able to guarantee that intersection does not exist. Nevertheless, the inverse does not hold true, as two objects may have empty intersection even though none of the six tests are true.

This technique is the one usually used to rapidly discriminate non-intersection between enclosing boxes, since it is easy to obtain the minimax parameters of a box that is described, for example, by one of its vertices and three vectors corresponding to the edges that intersect on this vertex. To obtain the maximum parameters it is sufficient to add to each coordinate of the reference vertex the corresponding positive componentes of the three vectors, and for the minimum parameters we will do the same with the negative ones.

Naturally, the minimax test can also be used to determine the absence of intersection between enclosing spheres, and, furthermore, in this case, it becomes easier to apply than the box case because of the following facts:

(1) If we describe a sphere by four parameters (x_0, y_0, z_0, R_0) the coordinates of its center and the value of its radius, then the minimax parameters are immediately obtained as

$$x_M = x_0 + R_0$$
$$x_m = x_0 - R_0$$
$$y_M = y_0 + R_0$$
$$y_m = y_0 - R_0$$
$$z_M = z_0 + R_0$$
$$z_m = z_0 - R_0$$

(2) If we consider the problem in the dynamic case, it is necessary to obtain the minimax parameters for each new object position. To do this, we need to find the new coordinates that define the container —box or sphere— in the new position. In the sphere case, this implies transforming the coordinates of a single point —the center— while in the box case, one vertex and three vectors must be transformed.

From the combination of both computations —the transforming of the container and the obtaining of the minimax parameters— it is easy to verify that, in the sphere case, the total computational cost comes to 24 elemental operations (18+6), whereas, for the box case we have a total of 45 operations (27+18).

If the minimax test fails —which is possible since it is not a necessary intersection condition— we will have to apply other tests to determine whether or not intersection occurs. In the sphere case, it is easiest to go directly to the exact equation which establishes that there is no intersection between two spheres B and C described by their parameters (x_B, y_B, z_B, R_B) and (x_C, y_C, z_C, R_C) respectively; that is, verify if

$$(x_B - x_C)^2 + (y_B - y_C)^2 + (z_B - z_C)^2 > (R_B + R_C)^2,$$

whose computational cost rises to 11 elemental operations.

In the enclosing box case different tests are usually applied when the minimax fails:

(1) The first alternative is to repeat the minimax test within a reference system with its planes parallel to the sides of the first box, and later in another with planes parallel to the second box. Obviously, this test is not a necessary interference condition either,

and the number of required operations is 124 (if only the first minimax test is applied) and 258 (if both are applied).

(2) Another possibility consists in verifying that for one of the twelve planes corresponding to the faces of the two boxes, all the vertices of one box belong to one of the half-spaces defined by the plane and all the vertices of the other box belong to a different half-space. This condition is sufficient, but it is still not necessary for no intersection. In the best cases this requires 109 operations, whereas, in the worst ones we will need 600.

(3) Finally, a necessary interference condition is obtained by checking if any edges of one object intersect a face of the other one.

In conclusion, we can affirm, beyond a doubt, that a sphere is always a better container than a box since, within the limits of the minimax test, the number of necessary elemental operations for defining minimax parameters is almost double in the box case even though the number of comparisons is the same in both cases (six). This difference is not important if we want to detect intersections between a set of n objects using an immediate algorithm $O(n^2)$, as these operations would not affect the quadratic term, and in practice, boxes and spheres will behave practically in the same way. When the minimax test fails, nevertheless, the situation radically changes since the number of necessary operations for making a comparison between two spheres determines the factor of n^2. We have seen that this number is 11 in order to have a necessary and sufficient non-intersection condition between spheres, whereas, in the box case this number is 109 in the best cases (for a single sufficient condition) and is more than 600 if we want to obtain a necessary condition. In equal circumstances, the complexity, then, comes down to being at least one order of magnitude greater for boxes than for spheres.

The confusion that has given rise to a false overvaluing of boxes as contrasted to spheres has been due to the fact that the 11 necessary operations for an *exact intersection test* between spheres are compared, on the one hand, with the six operations —at most— of the minimax test between boxes, on the other. The minimax is an approximate test and does not offer a necessary condition. This comparison, therefore, is in no way valid since both of the compared elements are not in equal conditions; an exact test is compared to an approximate method like the minimax. If the comparisons are carried out in equal conditions, the previous results, which clearly favor spheres, are obtained.

The only related question left to discuss is whether the minimax test will more frequently fail in one case than in the other, which is the same as asking oneself whether or not enclosing boxes are better than spheres as a way of representing the exterior shape of any object. In principle, it would seem that a box adapts better to the shape of an indeterminate object, since the fundamental property of spheres —isotropy— makes them more appropriate for representing objects in which their three dimensions —height, width and depth— are approximately of the same order, whereas with a box it does not matter whether the object is elongated (where one dimension predominates over the others) or flat (one dimension is notably diminished as compared to the others). Nevertheless, the application of the minimax test is equivalent to re-introducing a box or sphere into a new box with the property that its faces are parallel to the coordinate planes. In the spherical case, this box will always be the same regardless of the global reference system that is selected. In the box case, on the other hand, the second box may differ greatly from the first, to the extent that the position of the second is such that its faces form nearly 45° angles with the coordinate planes (Fig. 3.4), and will generally depend on the relative position of the object with respect to the coordinate axes that are taken.

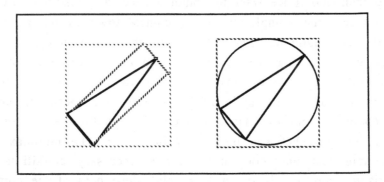

Figure 3.4   Fitting a box or sphere to the shape of an object

To conclude, as the question of adjustment to the real shape of the objects does not tip the scales towards boxes either, and taking into account the other numerous advantages of spheres that have been and will be mentioned, we can conclude that spheres are always preferable to boxes as a solid-body representation model.

b) Intersection Between Spheres in a Set.

Let us now move on to consider problem (2) for a set

$$S = \{ s_j, j = 1, 2, ..., n \}$$

in which each element s_j is a sphere that is completely described by its four parameters (x_j, y_j, z_j, R_j), i.e., center coordinates and radius. Our aim is to find out whether intersection exists between any two spheres belonging to this set, that is whether it is true that

$$\forall \ s_i, s_j \in S, s_i \cap s_j = \varnothing.$$

We have already mentioned that the particular case of a set of spheres is one of the few examples for which the complexity of the direct $O(n^2)$ algorithm could be reduced [Hopcroft, Schwartz and Sharir, 1983]. Concretely, the authors of this book know of only one other case: that of a set of boxes parallel to the coordinate planes [Edelsbrunner, 1982]. The algorithm is due to to Hopcroft, Schwartz and Sharir and it solves the intersection problem between a fixed set of spheres in $O(n\log^2 n)$ time. It is based on the *space-sweep* technique, a three-dimensional generalization of the well-known *plane-sweep* paradigm [Lee and Preparata, 1984]. We will present a summarized description of this technique, followed by a discussion.

To generalize the plane-sweep paradigm to space, we will logically have to carry out the sweep by use of a plane instead of a line. Let P be a horizontal plane that will intersect some spheres of a set S defining a set of circles C. An initial idea, the same as in the two-dimensional case, would be maintaining this set of circles C dynamically as P sweeps the space upward, checking with each change in C if an intersection between a pair of circles has taken place. This proposal, nevertheless, presents two problems: on the one hand, the radii of circles C continually vary with P, and on the other hand, it is difficult to maintain a set of circles in such a way that a fast intersection check is made easier.

The procedure, then, must necessarily complicate to include sweeps in various directions instead of just one. These directions must together form a network dense enough to guarantee that no intersection goes without being detected (the number of sweep orientations is independent of n). For each one of the sweep planes, we maintain a set of spheres C that intersect it in every moment, but we represent them by way of the maximum circle on the sphere which is parallel to the plane. This representation is simpler but it implies that a more complex mechanism must be used to detect intersections between the spheres of C. This approach may not detect some of the intersections, but if two spheres do intersect, the method guarantees that the intersection will be detected in at least one of the sweep orientations.

To define set Ψ made up of M sweep orientations, we must go to the next result. Let $s_i, s_j \in S$ be two spheres with centers c_i and c_j, and radius R_i and R_j ($R_i \geq R_j$), respectively. And let K be a cube inscribed in s_i and whose faces are parallel to the coordinate planes. Then, if angle θ between $c_i c_j$ and the line l parallel to z-axis and which passes through c_j is $\theta \leq \alpha$, where $\sin\alpha = 1/2\sqrt{3}$, then l intersects K. In accordance with this result, we will define Ψ by saying that for each orientation θ another orientation $\psi \in \Psi$ exists such that the angle between θ and ψ is less than or equal to α. The algorithm consists in applying the following spatial sweep technique for each $\theta \in \Psi$.

Let us suppose that θ points in the direction of the positive z-axis (to simplify the exposition). We sweep a plane P from $z = -\infty$ to $z = +\infty$. As P is being swept upward, we maintain a data structure T that contains some of those spheres of S whose centers are situated below P. Each time P passes through center c of a sphere s, we carry out the following two steps.

1. Let R be the radius of s. A range query is carried out which enumerates all spheres σ of T whose radius ρ is not greater than R and whose center is projected into a point P that belongs to the square of side $2R/\sqrt{3}$ with center on c. Each one of such spheres σ is

compared with s. If any intersection is found, the algorithm stops, otherwise sphere σ of T is eliminated.

2. Sphere s is added to T.

Clearly, the described procedure carries out n range queries and n insertions into T. Furthermore, each sphere $\sigma \in S$ may appear at most in one of those queries. On the other hand, the number of sweeps is a constant independent of n. It can be shown that the individual insertions and eliminations in T require time $O(\log^2 n)$, whereas each range query needs time $O(\log^2 n + t)$, where t is the number of spheres that fulfill the range condition of the query. Consequently, the use of this structure T gives rise to an algorithm that requires time $O(n \log^2 n)$.

Structure T is a balanced binary range tree with a certain amount of auxiliary data stored in its nodes. The leaves of T correspond to the spheres, sorted according to the x coordinates of their centers. Each internal node of T has an associated *priority search tree* (as is defined in [McCreight, 1985]). A formal proof of the correctness of the algorithm can be found in [Hopcroft, Schwartz and Sharir, 1983], to the effect that if an intersection exists, the algorithm is always able to detect it.

Next, we are going to compare this method with the immediate quadratic algorithm. When it is said that a complexity $O(n \log^2 n)$ is better than $O(n^2)$, naturally, we are referring to the asymptotic behavior of the algorithm, that is, for large values of n. In short, for this particular case it results that:

For $4 < n < 16$ $\rightarrow$ $n^2 < n \log^2 n,$

$n > 16$ $\rightarrow$ $n^2 > n \log^2 n.$

So that, without any more considerations, for values of n close to 10, a quadratic algorithm would be faster than the spatial sweep. Nevertheless, when studying the spherical representation in the exterior case, we have seen that in the most frequent case using around 10 spheres was enough in order to obtain an error of about 5%; that is, for us $n < 16$ most of the time and so the $O(n^2)$ algorithm will be preferable.

A more detailed study of the spatial sweep algorithm provides more arguments in favor of the quadratic method for our case. In *big-O* notation, usual in complexity theory, the constant factors in the dominant polynomial term are not included; notwithstanding, if we want to determine for which values of n an $O(n \log^2 n)$ algorithm is better than an $O(n^2)$ one, then we must take these factors into account. In the table below, the values of these terms are shown for different values of n when they are preceded by several different factors.

Factor M_1 for n^2 is easy to compute, since it is the number of operations that are necessary to determine whether or not two spheres have empty intersection. We have seen that if we directly apply the exact equation that compares the distance between the centers with the sum of the radii, the number of operations is 11. If we apply the minimax test beforehand, we would have 6 additional

n	n^2	$5\, n^2$	$11\, n^2$	$n \log^2 n$	$60\, n \log^2 n$
5	25	125	275	27	1.620
10	100	500	1100	110	6.621
20	400	2.000	4.400	372	22.320
50	2.500	12.500	27.500	1.592	95.520
75	5.625	28.125	61.875	2.910	174.600
100	10.000	50.000	110.000	4.413	264.780
200	40.000	200.000	440.000	11.684	701.040
300	90.000	450.000	990.000	20.314	1.218.840
500	250.000	1.250.000	2.750.000	40.192	2.411.558
750	562.000	2.812.500	6.187.500	68.412	4.104.754
1.000	1.000.000	5.000.000	11.000.000	99.317	5.959.011

comparisons. Therefore, in the worst case we would need 17 operations.

To estimate the value of factor M_2 for $n \log^2 n$ it is necessary to go back over the description of the spatial sweep method. On the one hand, the $\log^2 n$ factor is due to two contributions: the insertions into T and the range queries. As n queries and n insertions are carried out, we will have an initial factor with a value equal to 2. On the other hand, the whole operation must be repeated M times for each of M sweep orientations. Therefore, we can state that

$$M_2 = 2\,M$$

The value of M must be such that for each orientation $\theta \in \Psi$ there exists another orientation ψ of Ψ in such a way that the angle between θ and ψ is less than or equal to α, where

$$\alpha = \sin^{-1} 1/2\sqrt{3} = 0.293 \text{ rad.}$$

If we restrict the orientation set Ψ to a plane, we would have (Fig. 3.5) the minimum integer value for M given by

$$\pi / \alpha = 10.72 \quad \rightarrow \quad M = 11.$$

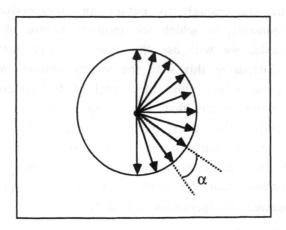

Figure 3.5 Set of orientations on the plane.

Therefore, to take in the three coordinate planes only, we would need a number of orientations of the order of 30, a value which establishes a lower bound for M. An upper bound is easily obtained if we take one of the coordinate planes and, for each one of the 11 orientations defined on it, we consider a plane with its 11 corresponding orientations so that with about 100 orientations, we would sweep all of the space with the constraint that for each orientation there must exist another such that the angle between them is less than or equal to α. Consequently, we can affirm that

$$30 < M < 100,$$

and including factor 2, we conclude that

$$60 < M_2 < 200.$$

If we now compare $17n^2$ with $60\, n \log^2 n$ we will conclude that we will have to deal with more than 200 spheres so that it is worth using the spatial sweep method (see table). Furthermore, while factor 60 is a lower bound for M_2, value 17 for M_1 is too high. Indeed, a cost of 17 operations for comparing two spheres would imply the performance, in each case, of 6 minimax test comparisons plus 11 for the exact test. None the less, it would be absurd to carry out all of these operations always, since if just one of them proves unfulfilled, that is enough to detect an intersection. For very cluttered environments, in which the frequent failure of the minimax test is foreseeable, we will use the exact equation that gives us a factor of 11. Comparing this with the sweep method, we see in the table that even for values close to a total of 500 spheres, the direct method gives better results than the sweep method.

We can refine this even more since, for not very cluttered environments, one of the minimax tests will suffice to detect an absence of intersection. If we assume that seven possibilities —each minimax comparison and the exact equation— are equally probable, the average number of operations would be

$$(1 + 2 + 3 + 4 + 5 + 6 + 17) / 7 = 5.43,$$

and if we only apply the minimax tests

$$(1 + 2 + 3 + 4 + 5 + 6) / 6 = 3.5.$$

As one of the minimax tests will usually be enough to determine whether there is intersection, it is valid to state that

$$3.5 \leq M_1 \leq 5.43.$$

We can take an approximate average value of 5 for M_1 without being too optimistic. In the table, we have tabulated the value $5n^2$, which compared with that of 60 $n \log^2 n$ tells us that for a total of $n = 1,000$ spheres the direct algorithm is still better than the spatial sweep method.

In addition to the above considerations, we must mention the fact that the factor of n^2 refers to elemental operations —adding, subtracting, products and comparisons— whereas we have not taken into account the numerous elemental operations that must be carried out in the sweep method including projecting each sphere on different sweep planes, ordering them in the direction of each sweep, etc. Thus, our result would be even more in favor of the direct method in the range of values of n that are of our interest and which will always be less than a thousand spheres.

3.2.2 Algorithm for Exterior Spheres

In the previous discussions we have only kept property (1) of our spherical representation in mind, that is, the fact that it is entirely made up of spheres. Now we will go on to consider the other specific properties of our hierarchical representation.

In the preceding section we have opted for a quadratic type algorithm that is based on comparing all possible pairs of spheres in search of an intersection. When posing the problem of how to improve the efficiency of an $O(n^2)$ algorithm we can think of two solutions:

1.- Decrease the size of the problem by reducing the number n of spheres involved.

2.- Reduce the number of operations by detecting as soon as possible the absence or existence of intersection. Since we are not dealing with the problems of counting or reporting intersections at the moment, we do not need to carry out all possible comparisons.

Property (iii) —the existence of a sphere hierarchy— will serve to apply the first of these solutions; while property (ii) — the existence of two sets of spheres— will be useful for applying the second solution. We will begin by analyzing the case of exterior spheres, after that we will consider the case of interior spheres and we will finish by proposing a final joint algorithm.

a) **Problem Definition**

Our problem may be defined in the following way. We want to determine the absence of intersection between two solids for which we have a representation hierarchy by means of exterior spheres. In accordance with our definition of exterior representation, we can guarantee that there will be no intersection between two objects if there is no intersection between the spheres that make up their respective exterior representations. Our problem consists in improving the two object representations more and more until we reach a moment in which the intersection between both is empty, that is

$$\exists \; n, m \; ; \; \forall \; S_{Bi} \in S_B(n), \; S_{Cj} \in S_C(m), \; S_{Bi} \cap S_{Cj} = \emptyset;$$

where n and m denote the order number assigned to representations $S_B(n)$ and $S_C(m)$ for B and C respectively.

The method is valid because if there is no intersection between the spheres of the representation, there will be no intersection between the boundaries of the two solids, since the spheres define a covering for such boundaries. The rare case in which an object is entirely contained inside another can be immediately detected since the initial representations with a single sphere make up a complete covering of the object, and one of the spheres will be contained in the other, at least partially. Then, this case will be given specific treatment which include the use of interior spheres.

It must be noted that, from the intersection detection point of view, we have before us a case that corresponds to problem (3) with two sets of spheres. Indeed, we do not want to find the intersections between all the spheres of a set, rather, we want to determine, for two fixed representations, if there is no intersection between any two spheres belonging to two different sets. This is a further reason for not employing the spatial sweep method which only serves to solve problem (2).

Therefore, the basic starting algorithm will be $O(n_B n_C)$, where n_B and n_C are the number of spheres that make up each one of the two exterior representations for objects B and C, and it is to our advantage to make these numbers as small as possible.

b) Description of the Algorithm

In accordance with the exterior representation construction method described in the preceding chapter, we can define a set E made up of all spheres that take part in any of the representations of the object. The number of this set's elements will depend on the number of representations that have been defined so far for the object. So, to denote a certain representation $S(n)$, we can say that the spheres that make it up are *active* spheres within set E, whereas the rest of the spheres are *passive*. When we improve a representation and some spheres are substituted by others, we can say that some spheres have been *activated*, whereas others have gone back to being passive. By construction, once a sphere has gone from active to passive, it cannot go back to being activated.

The algorithm for determining the absence of intersection between two objects B and C proceeds, then, in the following way. We start with two representations with a single sphere for each object S_{B1} and S_{C1}, so that if

$$S_{B1} \cap S_{C1} = \varnothing$$

we have finished and we can affirm that there is no intersection between the two objects. If this is not so, we *mark* each one of these two spheres and moreover, we label them as *old*. After that, we improve the *worst* representation, according to the value of δ

coefficient (let us suppose that it is that of B), until S_{Bl} stops being active. All of the spheres that have gone active are also labeled, this time as *new*. At that moment we compare all of the new spheres of E_B with all of the active spheres of E_C (only S_{Cl} for now). Each time that a new sphere has been compared with all of the active spheres of the other set, it is labeled as old. If we detect no intersection at all, we have finished; if not, then let us suppose that we detect

$$S_{Bi} \cap S_{Cj} \neq \emptyset.$$

Then, we mark the two spheres S_{Bi} and S_{Cj}, we improve the worst representation by adding new spheres that will substitute the other ones until the marked sphere stops being active, and we begin again by comparing all of the new spheres of the improved representation with all of the active spheres (new and old) of the other one. If we find no intersection, we will compare all of the new spheres of the other set (which could have been left over) with all of the old spheres of the first one, so that we go through all of the possible pairs of spheres of the two sets.

In Figure 3.6 sets E_B and E_C corresponding to objects B and C are depicted as two lists having N_B and N_C as lengths. Active spheres

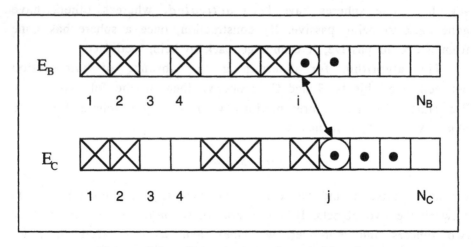

Figure 3.6   Data structures for exterior spheres

are shown by means of the symbol (×), and those which are not only active but are new as well are indicated by a dot (•). We have identified the two intersecting spheres by *marking* them with a circle. Therefore, the ×'s tell us which ones are the spheres that take part in the representation in a given moment; and once we have improved a representation to the point where the marked sphere is eliminated, we will have to compare this representation's spheres that are indicated by a dot (the new ones) with all of the other representation's active ones (indicated by a point or an ×). Each time a new sphere is compared with all of the active ones of the other set, we substitute an × for a dot to indicate that it is no longer new.

By means of this technique, it is obvious that what we are trying to do is to avoid redundant comparisons between the same spheres. In this way, we know that an *old* sphere has already been compared with all of the old spheres of the other object, therefore, we will only compare it with new spheres.

To avoid those situations where we refine a representation many times without managing to eliminate the marked sphere, —which will happen when the particular sphere is very well adapted to the external shape of the object— we will only improve the worst representation until its quality coefficient becomes less than that of the other one, and therefore, it then becomes the best. That is, in each step, we check which is the worst of the two representations at that moment in order to improve it, then we will continue in this way until one of the two marked spheres stops being active.

It is important to call our attention to the fact that the present algorithm will never be able to detect an intersection, rather, it only guarantees the absence of intersection if it does not exist. This is due to the very nature of the exterior sphere representation. It can be said that the algorithm is complete for a given resolution, in the sense that if two objects have no intersection, we will be able to affirm it provided that we use an accurate enough level of representation.

The algorithm will behave well in what regards to its termination. It will always end either at the moment that it detects intersection absence, or when the lowest available level of representation is reached. Naturally, it will give only a solution in the first case.

Regarding the complexity of the algorithm, we must say that an upper bound in the worst cases will be given by $O(N_B N_C)$. It is evident that this estimation is excessively pessimistic since:

1.- There is no case in which all the elements of E_B are compared with those of E_C, rather, only the active spheres of each set are compared.

2.- In most non-intersection cases, a few spheres from each set will be enough to determine intersection absence.

Therefore, the expected average complexity will be the result of assigning to N_B and N_C values close to 10. Note that the values of N_B and N_C depend on the maximum levels of representation that we have defined. These levels should be as numerous as possible in order to be able to discriminate difficult cases. In the average case, however, we will only use a very small portion of available spheres, since we only employ the spheres that are necessary for determining non-existence of intersection. In this way we have fulfilled one of our proposed objectives, namely, to decrease as much as possible the size of the problem so that with a few spheres we can get a solution.

3.2.3 Algorithm for Interior Spheres

The representation using interior spheres can be considered complementary to the representation with exterior spheres, in the sense that while the latter has served to determine lack of intersection, interior spheres are going to be useful for detecting existence of intersection.

Indeed, if there exists intersection between two interior spheres belonging to the representations of two different objects, we can state that the intersection between the two objects is not empty. This is so due to the fact that in accordance with the definition of interior representation, all of the points that belong to an interior sphere lie in the interior of the object that they represent. In this way, points common to two inner spheres corresponding to two different objects, will also be common to both solids, and will belong to the intersection between both, which, therefore, will not be empty.

The algorithm for this case is similar to the one for outer spheres, although the nature of interior spheres allows us to make some simplifications. In the first place, we look for a single intersecting pair of interior spheres:

$$\exists \, S_{Bi} \in I_B, \; S_{Cj} \in I_C \; ; \; S_{Bi} \cap S_{Cj} \neq \varnothing;$$

where I_B and I_C are the sets composed of all spheres that take part in any of the representations of each object. Finding one such pair is sufficient for the problem to be solved. Like in the previous case, we are dealing with an instance of problem (3) for two sets of spheres corresponding to each pair of representations that we compare.

On the other hand, when we studied the interior representation we saw that we constructed new representations by adding new spheres to the ones we already had —instead of substituting some for others as we did in the exterior case—, this means that an *active* sphere will never stop being active. In addition, the distinction between *old* and *new* spheres will interest more than the assigning of each active sphere to one of the different representations.

The algorithm will follow the steps that we shall now describe. Initially, we will have the first two representations formed by two sets of active new spheres. We will compare all of the pairs of spheres from the two sets, one with another, in search of intersection; if we find one pair that intersect each other, our problem comes to an end. If this is not the case, we would label all of the spheres from the two sets as *old* and we would add new spheres to the worst representation. Following that, we would compare all of the new spheres with the spheres from the other set. If an intersection is detected we would have finished; otherwise, the new spheres would become *old* and we would repeat the process.

The only point which has not been clarified would be determining how many new spheres would have to be added to the worst representation in each step of the algorithm. The most reasonable criterion is to use the incremental technique of interior representation construction directly, in such a way that we bring forward the group of additional spheres that define the next representation. On the other hand, to decide which of the two representations is worse it is natural to use the parameters (δ or ε) introduced when we described the process of improving an interior representation.

With regard to this algorithm's properties, we can come to conclusions similar to the exterior case. In the first place, it is evident that with this algorithm we will never be able to determine the lack of intersection between the two represented objects if the two objects, in fact, do not intersect. In any case, this is obviously not our goal. The algorithm is complete for a given resolution; that is, an existing intersection will always be detected provided that a sufficient number of interior spheres are used. The algorithm will always end up either by having reached a solution or by having surpassed the deepest level in the hierarchy of representations, which has been fixed beforehand.

The complexity in the worst case will also be $O(N_B N_C)$, that is quadratic in the number of elements of sets I_B and I_C. In this case, we can also state that the average complexity case will greatly differ from the worst case. This is so because if there is intersection between the objects, we will usually need no more than a few dozen spheres in each representation even though we may have large N_B and N_C values in order to discern difficult cases.

3.2.4 General Algorithm

The moment has come to propose a general algorithm to solve the intersection detection problem. It will be general because it combines the use of exterior and interior representations, and because it will serve to detect intersections between several objects.

a) Algorithm for two objects

Up to now we have made use of property (iii) of our representation —which refers to the existence of a hierarchy of spheres— in order to succeed in reducing the size of our problem to the maximum. In the simplest case, then, a couple of spheres would be enough to solve it. Now we have to unify the two proposed

algorithms in order to apply property (ii) of the spherical model
—representation duplicity— to decrease the number of operations,
detecting intersection absence or existence as soon as possible.

We have previously seen that each one of the two proposed
algorithms had an inherent limitation that was due to the nature of
the representation used in each case. On the one hand, using exterior
representations, it was impossible to detect intersection existence
between two objects if this intersection was indeed happening; and
on the other hand, with interior spheres we were not able to
guarantee that two objects did not intersect when in fact they did
not. These limitations lead to absurd situations in which, for example,
when we have two objects that overlap each other, we would apply
the algorithm for exterior spheres over and over again until we
exhaust our representation capacity without ever coming to a
conclusion and after having used a high number of spheres.

By means of the simultaneous and joint application of the two
representations, we will succeed in avoiding situations like this one
and we will obtain an algorithm which will produce good results in
any case. We assume that two sets of spheres, SE_B and SI_B, are
available for object B, the first one corresponding to its exterior
representation and the second one to its interior representation; and
similarly for object C. Then, testing for intersection among these sets
may result in one of three possible situations:

(1) $SE_B \cap SE_C = \emptyset \rightarrow B \cap C = \emptyset$.

(2) $SI_B \cap SI_C \neq \emptyset \rightarrow B \cap C \neq \emptyset$.

(3) $SE_B \cap SE_C \neq \emptyset$ and $SI_B \cap SI_C = \emptyset \rightarrow B \cap C = ?$.

In the first case, it can obviously be guaranteed that the two
objects do not intersect. Similarly, in the second case it can be
assured that the two objects do intersect. In the third case, however,
we are lacking information to decide whether an intersection occurs
or not; our only alternative to solve this indeterminacy is to improve
the representations for the objects and test again.

A heuristic procedure is used to decide which representation
must be made better and how much it must be improved before the
test for intersection is executed once more. First only outer spheres

are used, starting with the highest level representation —o n e
enclosing sphere per object— we improve them trying to check if
situation (1) is true (for this purpose the especial algorithm for
exterior spheres is used). Up to this point, we have assumed an
absence of intersection and tried to prove it. When a *first threshold*
is reached (given by the first local minimum for δ_{tot} quality
coefficient) we suspect that our initial assumption may be wrong and
that we are in a situation of *possible intersection*. To verify it we
check if situation (2) is true by using only inner spheres until a
second threshold is reached. Now, if the indeterminacy persists after
this second threshold, we say we are facing a *hard case*. To remove
the indeterminacy a cross-test is made for outer and inner sets of
spheres. This cross-test may yield one of four possible results:

(4) $SE_B \cap SI_C = \emptyset$ and $SI_B \cap SE_C = \emptyset$.
(5) $SE_B \cap SI_C \neq \emptyset$ and $SI_B \cap SE_C \neq \emptyset$.
(6) $SE_B \cap SI_C \neq \emptyset$ and $SI_B \cap SE_C = \emptyset$.
(7) $SE_B \cap SI_C = \emptyset$ and $SI_B \cap SE_C \neq \emptyset$.

In the first case, when intersection between exterior and interior
spheres does not exist, we have the almost evident *suspicion* that
there is no intersection between the two objects and that we must
improve the exterior representation in order to show it.

In the second case, the interior and exterior spheres of the two
objects intersect one another, which is a symptom of a very probable
intersection between the objects. Consequently, we will proceed by
improving the interior representations so that it will be detected.

The third case does not seem so clear cut, as we apparently have
contradictory information. On the one hand, the fact that $SE_B \cap SI_C \neq \emptyset$ seems to indicate that indeed there is intersection while $SI_B \cap SE_C = \emptyset$ would incline us to believe that there is lack of intersection. To put
an end to this indeterminacy, we improve SI_B and SE_B. In the first
case we suppose that there is intersection and that the second
equality is misleading. In the second case, however, we suppose that
there is no intersection and that the first inequality is the one that is
incorrect.

Finally, in the fourth assumption, following an identical line of reasoning as above, we would draw the conclusion that the right decision is to improve SI_C and SE_C.

The algorithm will follow a formal scheme equivalent to considering that the four previous cases correspond to the conditions of four rules, so that one of them has an associated action. Each time we apply one of these actions we would go through the rules again to determine the one to be applied — since they are mutually excluding— and then we would start over again.

We must note that the algorithm's success in the *hard* case depends on the fact that the two preliminary thresholds are adequately fixed.

To guarantee that the algorithm behaves well, it is a good idea to define a higher-level strategy — in the way of *meta-rules*— that serves to detect unusual cases and adequately modify the strategy. These meta-rules carry out a tracking of the frequency and sequence of the application of the four previous rules in such a way that if, for example, it is detected that the first rule is successively applied with no success, we would have a clear indication that the hypothesis of non-intersection is incorrect and we would go on to improve the interior representations.

Finally, we should draw our attention to the fact that when we carry out crossed checks between interior and exterior representations, we must slightly complicate data structures in order to implement them. In particular, in order to avoid repeated comparisons, a single bit or flag (old/new) for each sphere will not be enough. Now we will have to use two bits, one which indicates if the sphere has been compared with all of the active spheres of the other object's interior representation, and another one that indicates if it has been compared with the exterior representation.

3.3 Moving Objects

Spheres prove to be even more adequate to deal with the general collision detection problem, i.e., when some objects are actually moving.

We could directly apply the multiple intersection test to our case with no additional considerations since we have algorithms to detect intersections between spheres. Nevertheless, we are going to discard this possibility. The fundamental reason has a theoretical nature and it will only be fully understood when we apply collision detection to motion planning in the next chapter. The problem lies in the fact that to generate safe paths (collision-free), we have to guarantee the lack of intersection in any of the infinite points that make up these paths in Configuration Space. This is not possible by means of the multiple intersection test —as it is intrinsically discrete— unless we make estimations regarding minimal distances and maximal velocities. As we shall see, we want to obtain *safe paths* independently of the velocity at which they are traveled. This makes these estimations pointless.

The well-known sweeping method has been selected for our application. Its main difficulty is the computation of the actual swept volume in a general case, but this drawback is eliminated by using the spherical model, since in the particular case of a sphere that computation turns out to be, again, a simple matter [de Pennington, Bloor and Balila, 1983]. Indeed, the volume swept by a moving sphere can be specified by the segment of a curve described by its center (Fig. 3.7). If $\mathbf{v_B} = (x_B, y_B, z_B)$ is a vector describing the position of the center of a sphere B, then the locus of positions of the center of B for a certain motion will be a curve as given by its parametric equation: $\mathbf{v_B} = \mathbf{v_B}(s)$, $s \in I$, where I is a real interval. Then, the swept volume can be defined as:

$$SVB = \{ \, Q \, ; \, \exists \, s \in I \mid d(Q, \mathbf{v_B}(s)) \leq R_B \, \}.$$

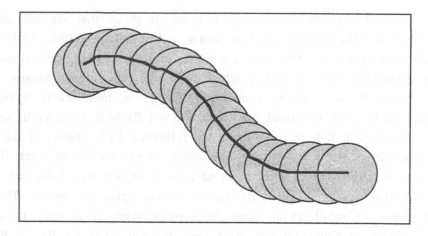

Figure 3.7　　Volume swept by a moving sphere

Consequently, a condition for absence of collision between two moving spheres is simply given by:

$$\forall\, t \in [t_0,\, t_f],\ \ d(\mathbf{v_B}(t),\, \mathbf{v_C}(t)) > R_B + R_C,$$

where time has been used as the parameter of the curves. This condition is equivalent to showing that the minimum distance between two curve segments is greater than the sum of the radii of the spheres.

As a result, to assure that two objects in motion do not come into collision, the previous algorithm for intersection detection can be applied with just a slight modification: instead of testing spheres for intersection, we test their corresponding swept volumes for intersection, which amounts to computing upper bounds for distances between curve segments.

3.4 Application to a Manipulator

The previous results can be applied to determine whether the motion of a robot manipulator is free of collisions. To define the scope of this problem certain assumptions must be made. First, we assume that all the obstacles in the environment of the robot are at rest. Second, we consider only two types of joints that are the most widely used for robots: revolute joints and prismatic joints. And the third assumption is that the complete motion can be decomposed into *elemental motions* during which only one robot joint changes.

Now, if we define a manipulator as a mechanical system composed of a set of joined rigid bodies $B = \{ B_i, i = 0, 1, ..., n \}$ in such a way that each link B_{i-1} is tied to the following B_i by means of the i-th joint (B_0 is the fixed base of the robot). And in addition, each link B_i can be partitioned to yield a set of convex objects $B_i = \{ B_{j_i}, j = 1, 2, ..., n_i \}$ that are represented by means of the spherical model. Then, regarding the robot as an open-chain mechanism, it is easy to see that, during an elemental motion of the i-th joint, links $B_0, B_1, ..., B_{i-1}$ do not move, while links $B_i, B_{i+1}, ..., B_n$ move together as if they were all part of a unique rigid body.

Therefore, our problem has been reduced to the determination of the distance between curves and fixed points; the curves are described by the centers of the spheres representing the moving links, and the fixed points are just the centers of the spheres representing the links that do not move, as well as the obstacles. In addition, the resulting curves are simple enough: arcs of circumference for a revolute joint, and line segments for a prismatic joint.

Consequently, to detect a collision during the motion of a manipulator, we have to solve two simple geometric problems: computing the distance D between a point and a line segment, and computing the distance D between a point and an arc of circumference. In Fig. 3.8 both situations are shown; in (a) q_i represents the generalized coordinate for a revolute joint and in (b) q_j is the coordinate for a prismatic joint. In each case an elemental motion is given by a certain increment Δ in the joint coordinate.

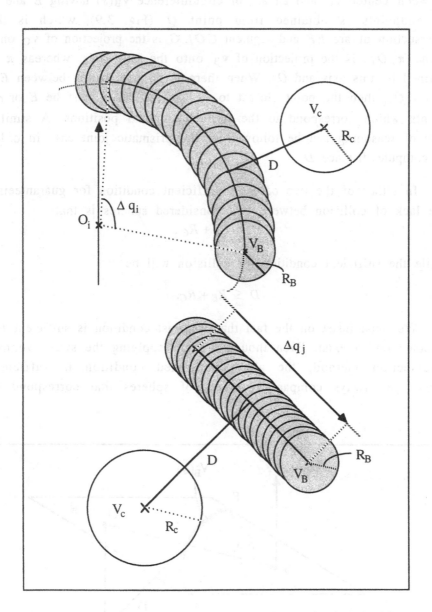

Figure 3.8   Detecting collisions for revolute and prismatic joints

In the concrete case of the revolute joint, the minimal distance D between center $\mathbf{v_C}$ and an arc of circumference $\mathbf{v_B}(s)$ having E and F as endpoints is obtained from point Q (Fig. 3.9), which is the intersection of arc EF and segment GO_i. G is the projection of $\mathbf{v_C}$ onto plane π; O_i is the projection of $\mathbf{v_B}$ onto the joint axis, whereas π is defined by this axis and O_i. When there is no intersection between EF and GO_i, then the point closest to $\mathbf{v_C}$ on the curve must be E or F, points which correspond to the initial and final positions. A similar line of reasoning can be followed in the prismatic joint case in order to compute distance D.

In either of the two cases, a sufficient condition for guaranteeing the lack of collision between two considered spheres is that
$$D > R_B + R_C ,$$

while the sufficient condition for collision will be

$$D \leq R_B + R_C.$$

We must insist on the fact that this last condition is sufficient for intersection to exist. Even though we are applying the swept volume intersection method, the above-mentioned condition is sufficient since we always compare two sets of spheres that correspond to

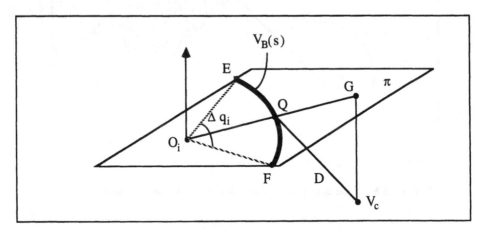

Figure 3.9   Minimum distance between a point and a circle arc

objects that are not simultaneously in motion.

As a conclusion, we have solved the problem of computing the distance between the curves described by the centers of some spheres and the centers of others, for our particular case. In consequence, we can directly apply our previous conclusions regarding collision detection between objects to the case in which the moving objects are part of a robot's links.

In summary, the algorithm for detecting collisions in a robot's given motion will go through the following steps.

Let a robot be defined as a mechanical system made up of a set of linked rigid bodies

$$B = \{ B_i, i = 1, 2, ..., n \},$$

in turn, for each solid B_i a partition in $n^i{}_b$ convex objects apt for being represented in accordance with our spherical model will be defined.

$$B_i = \{ Bj_i, j = 1, 2, ..., n_i \}.$$

On the other hand, we will have obstacle set C which is made up of all fixed rigid bodies situated in the robot's environment

$$C = \{ C_r, r = 1, 2, ..., m \}.$$

For each object C_r a partition into $n^r{}_c$ convex objects that can be *spherized* will be defined.

$$C_r = \{ Cs_r, s = 1, 2, ..., m_s \}.$$

For each one of these objects Bj_i and Cs_r we will define a double hierarchy of spheres in accordance with our spherical representation model.

Then, for a particular elemental robot motion corresponding to the i-th joint, we will divide all of the objects into two sets to which we will assign different *colors*. On the one hand, we will have objects that are a part of solids $B_1, B_2, ..., B_{i-1}$ and those of C. On the other hand, we will have objects in $B_i, B_{i+1}, ..., B_n$; these being the only ones in motion and moreover, no relative motion will exist between them. Next, we will compare all of the differently colored objects (except obviously B_{i-1} and B_i) in search of a collision. To do this, it will be necessary to compute the paths —straight segments or arcs of circumference— that describe each center of each sphere of the different representations that are used for each object. This problem is easy to solve since we can find the coordinates that give us the initial position of the centers from the transformation matrix of an object corresponding to the initial position of an elemental motion, and then, according to the type of joint and the corresponding coordinate increment we completely define the path. Note that the motion is identical for all points: the same rotation or the same translation.

In general, a few spheres will be enough to decide whether or not a collision takes place and we will only have to turn to the algorithm for the so-called hard cases when we are faced with a tight fitting motion that comes close to contact, or with a slight touch with mutual penetration.

Chapter 4

Motion Planning

We are now going to consider the use of our spatial representation model for the solution of the general problem of motion planning of robot manipulators in three dimensions. The algorithm that we will propose is supported, to a great extent, by the results of the preceding chapter, since collision detection, as we shall see, is a preliminary problem for being able to find obstacle-free paths.

The method we present does not try to exhaust the application of our spatial representation as a useful tool for motion planning; rather, it should be taken as an initial approach with the hope and certainty that in the future many algorithms based on the simplicity, elegance and versatility of our spherical model, will have taken its place.

In this chapter we will begin by rigorously defining the problem, discussing the influence of the different aspects involved. We will also consider the possible variants of the general problem. Afterwards, the proposed algorithm will be presented which is based on the spherical representation. To conclude, we will show the results that correspond to an application example for a robot that moves in a particular environment.

Chapter 4

Motion Planning

We are now going to consider the use of our spatial representation model for the solution of the general problem of motion planning of robot manipulators in three dimensions. The algorithm that we will propose is supported, to a great extent, by the results of the preceding chapter, since collision detection, as we shall see, is a preliminary problem for being able to find obstacle-free paths.

The method we present does not try to exhaust the application of our spatial representation as a useful tool for motion planning, rather, it should be taken as an initial approach with the hope and certainty that in the future, many algorithms based on the simplicity, elegance and versatility of our spherical model, will have taken its place.

In this chapter we will begin by rigorously defining the problem, discussing the influence of the different aspects involved. We will also consider the possible variants of the general problem. Afterwards, the proposed algorithm will be presented which is based on the spherical representation. To conclude, we will show the results that correspond to an application example for a robot that moves in a particular environment.

4.1 Statement of the Problem

We are going to rigorously present the question of motion planning, the name with which we designate the set of problems related to the path computation for a robot. In the last few years, the size of this set of problems has grown due to the interest that has arisen in many researchers who have incorporated new variants for the original problem. In this study, we focus on the case of a robot manipulator that moves in ordinary three-dimensional space having as goal a specified final configuration. In its movement, the robot wil avoid colliding with the surrounding objects that will be known a priori. Another fundamental area of research corresponds to the study of the movement of mobile robots, which make up the other large group in which robots can be classified.

4.1.1 Preliminary Concepts and Terminology

In the following we define a series of concepts and ternimology which will be used later on.

Real Space. This is the physical space in which our robot is immersed. We will suppose that it is the Euclidean two or three-dimensional space and that it is solely occupied by rigid bodies.

Environment E. This is a portion of the real space that encompasses all the positions that the robot can physically reach, that is, it must include the robot's workspace in the case of a fixed-base manipulator (for an interesting discussion on robot reach see, Korein, 1985]), or either in the mobile robot case or that of a single moving object, it must include the portion of real space which confines it. Logically, E will have to be a closed subset of real space and will contain all the obstacles that will have to be taken into account when avoiding collisions.

Robot B. We consider a robot to be a mechanical system formed by a set of n_b rigid bodies that can be linked together:

$$B = \{ B_i, \; i= 1, 2, \; ..., \; n_b \}$$

In turn, for each solid B_i a partition into $n^i{}_b$ convex objects will be defined so that it will be valid to write:

$$B_i = \{ B_{j_i}, \; j=1, 2, \; ..., \; n^i{}_b \}$$

Configuration. Like all rigid body systems, a robot is characterized by a certain number n of degrees of freedom (dof), that specify the minimum number of parameters that define a spatial position for the robot. Thus, in the case of holonomic joints, a configuration of the robot will be given by n values corresponding to n generalized independent coordinates:

$$(q_1, \; q_2, \; ..., \; q_n).$$

The physical meaning of these parameters will habitually be that of distances or angles, although another type of magnitude would be admissible.

Placement. To refer to the set of solids that make up robot B situated in positions that correspond to a given configuration $(q_1, q_2, ..., q_n)$ we will use the term *placement* and we will denote this by

$$B \, (q_1, q_2, \; ..., \; q_n)$$

or, in abbreviated form,

$$B \, (q_k), \; k = 1, 2, \; ..., \; n,$$

and for a particular solid:

$$B_i \, (q_k) \quad \text{or} \quad B_{j_i} (q_k)$$

Obstacle Set C. We will call the set composed of all those objects not included in B and situated in the environment E, the *obstacle set* C. Although each individual obstacle can be defined arbitrarily, we will admit each obstacle as a connected component of E.

$$C = \{ C_r, \; r= 1, 2, \; ..., \; n_c \}.$$

In turn, for each obstacle C_r a partition into $n^r{}_c$ convex objects will be defined in such a way that it is valid to write

$$C_r = \{\ C j_r, j=1, 2, ..., n^r{}_c\}.$$

Safe Configuration. We will say that a certain robot configuration (q_k) is *safe* or that it corresponds to a *free placement* if it verifies that

$$B\ (q_k) \cap C = \varnothing.$$

That is, there is no intersection between the robot and any of the obstacles for the given configuration. A *semi-free placement* is often defined as one in which the previous equality is verified if instead of C, we take the interior of C; in other words, contact between robot and obstacle is allowed, but mutual penetration is not.

Free Space *FP*. Free Placement Space *FP* or simply Free Space is the set formed by all safe configurations

$$FP = \{\ B\ (q_k)\ |\ (q_k) \in Q \text{ and } B\ (q_k) \cap C = \varnothing\ \}$$

where (q_k) goes over the subset of *admissible configurations* $Q \subseteq \Re^n$, according to the physical limits of each generalized coordinate.

Motion. To rigorously define a motion or *path* μ for robot B between two placements $B(q^S{}_k)$ and $B(q^G{}_k)$ we will say it is a continuous mapping μ of the unit interval $[0, 1]$ in free space:

$$\mu : [0, 1] \to FP, \qquad \text{such that} \qquad \mu(0) = B(q^S{}_k), \qquad \mu(1) = B(q^G{}_k).$$

For simplicity, instead of using a parameter s whose definition domain were a certain interval I of $\Re$, we have *normalized* I to the be the unit interval. Also note that we assign no physical meaning —time, for instance— to this parameter. Moreover, we observe that in this definition the concept of *safe* motion is implicit, that is, an admissible motion can be regarded as a continuous sequence of safe configurations, since it has been defined as a mapping in free space.

Margin. Introducing the usual Euclidean metrics in real space, we can provide free space with a distance function d between two points M and N, so that we define the *margin* of a certain placement of robot $B(q_k)$ as:

$$margin\ (B\ (q_k)) = inf\ \{\ d\ (M,\ N);\ M \in B\ (q_k),\ N \in C\ \}$$

That is, the margin would be the minimum distance from the robot to the nearest obstacle. Likewise, the margin for motion μ would be:

$$margin\ (\mu) = inf\ \{\ margin\ (\mu(s));\ s \in [0,\ 1]\ \}$$

Configuration Space CS. This is defined as a subspace of $\Re^n$ so that each point of this space corresponds to an admissible configuration $(q_1, q_2, ..., q_n)$. Even though by definition, the set of admissible configurations Q will coincide with space CS, it may not be so; for if some non-holonomic joint exists, Q could not be parameterized by means of generalized coordinates. Configuration space presents the evident advantage of being able to bring together the geometric and kinematic aspects of the problem in a single description.

Several of the preceding definitions can be extended to Configuration Space, so that we have:

Obstacle. For each obstacle $C_r \subset E$, an obstacle CSC_r in Configuration Space will be defined as

$$CSC_r = \{\ (q_k);\ B\ (q_k) \cap C_r \neq \varnothing\ \}.$$

Free Space. An initial definition could be

$$CSFP = \{\ (q_k);\ B\ (q_k) \cap C = \varnothing\ \}.$$

Nevertheless, it is more rigorous to define it in terms of margin as

$$CSFP\ (\varepsilon) = \{\ (q_k);\ margin\ (B\ (q_k)) \geq \varepsilon\ \},$$

$$CSFP = \lim_{\varepsilon \to 0} CSFP(\varepsilon)$$

We take the closure of this set in the semi-free space case. We have stressed making a clear distinction between free space (FP) as a subset of real physical space on the one hand, and free space $CSFP$ as a subset of configuration space on the other. This distinction is sometimes unclear which may lead to confusion.

We should also emphasize that obstacles and Free Space *fill up* Configuration Space, which is not so for real space with the exception of some trivial cases. Free space FP should never be confused with the complementary set of the obstacle set C.

Motion. This is a continuous curve in $CSFP$ that joins the two points (q^{S_k}) and (q^{G_k}) that represent the initial and final configurations.

4.1.2 Taxonomy of the General Problem

We will generically define the *general motion planning problem* as:

Given a robot B and an environment E occupied by a set of obstacles C, find a motion μ for B among the obstacles in C that fulfills certain given conditions and is optimal in a certain sense.

The different families of concrete problems that arise from this general definition differ in the nature of the elements involved in each one. We can propose the following taxonomy with these differences in mind.

Dimension of Real Space. This is a fundamental factor when posing the problem. The problem becomes drastically simplified if we reduce it to the plane. From there, the number of proposed approaches to the three-dimensional case will be notably inferior compared to the two-dimensional case.

Complexity of the Environment. Sometimes, the particular case of motion in a very limited space that remains fixed is studied. We will call it a *local planning* problem and we will use the term *local expert* to refer to the planner. With this, we try to take on very concrete problems such as turning in a corridor or going through a door whose solution can later be used as part of an algorithm for a more widely-ranged motion.

Physical Nature of the Robot. In general, we can distinguish between two fundamental cases. In the first place a manipulator robot can be considered from the kinematic point of view, as an open-chain mechanism, so that B_0 would be a fixed or base element of the robot, rigidly joined to the floor, and each kinematic element B_i would be linked to the previous element and to the following one, if one exists (the last element has no successor). Moreover, each element would be made up of a certain set of convex objects rigidly joined together. The joints between the different elements will be the usual ones: basically, revolution and prismatic joints; without discarding others like spherical, cylindrical, etc.

The second case is that in which $n_b = 1$; that is, the robot is made up of a single rigid body that, logically, is not fixed to the floor. Sometimes this is called a *rigid robot*, even though when referring to the planning of its motion other names are used like the *piano movers' problem* and, if the real space has two dimensions, it is sometimes called a *mobile* robot. The authors of this book believe, however, that this latter term is not correctly applied in this context since, as it will be argued further on, the kinematic constraints to which a mobile robot is subjected, makes its motion unable to resemble that of the rigid body on the plane. For three-dimensional space, the term *flying object* has also been used to refer to this case. In order to unify criteria we will call a rigid body with no linking that moves in real two or three-dimensional space a *piano*. One particular case is to reduce the piano to a single point.

Nature and Dimensions of Configuration Space. The construction and characterization of *CS* is not at all an easy task. This problem by large, is conditioned by the nature of the robot. In this way, the robot manipulator case is notoriously more complex than the piano case.

Moreover, the greater the number of the robot's degrees of freedom, the more difficult the problem will become, as the dimensions of *CS* are increased. In the piano case only two possibilities exist: 3 dof for the plane and 6 dof for real three-dimensional space, although motions are sometimes simplified by restricting them to simple translations, for example. A robot manipulator can have an indefinite number of dof; although we speak of *redundant* robots for a number greater than 6, since more than one possible configuration can exist to place the robot's hand in a desired position and orientation.

From the topological point of view, we can formulate the problem in the following terms: this consists in finding a connected component in $CSFP$ space that contains the initial and final configurations. This theoretical statement is of little practical interest, since, as we have already stated, it deals with a space characterized by a high number of dimensions and with irregular boundaries, which makes efficient computation difficult.

Object Representation. We must keep in mind the kind of objects that can be represented, as well as the method used to represent them, not only for the obstacles, but also for the robot links. In general, these two factors somehow tend to be limited: either because any body involved in the problem must belong to a certain set of objects, or because it must adjust itself for certain criteria in order to be represented.

Motion Conditions. The most usual case is that in which only the start (q^{S}_{k}) and goal (q^{G}_{k}) configurations are specified, motion being restricted to beginning at the start configuration and ending at the goal one. Note that according to the previously given definition, the idea of avoiding collisions with objects is implicit to the concept of motion, as motion takes place in free space. Sometimes, the goal configuration is substituted by a certain set of positions in such a way that any motion that joins the start position to any of the configurations of this set would be valid.

Motion Optimization. The demand for optimal motion has only been thoroughly handled in the case where the robot resembles a physical point, or in other situations that can easily be reduced to this one. In these cases, we will look for the shortest Euclidean path that fulfills the motion conditions. In more complicated cases, the very complexity of the problem makes us limit ourselves to looking for solutions which are *acceptable* to a certain extent rather than being *optimal*. The problem is then to develop techniques to find these acceptable solutions. A criterion that has sometimes been proposed is for motion to maximize margin in order to guarantee, in this way, that the robot does not come too close to the obstacles. Nevertheless, this criterion alone can give rise to trajectories that tend *to go around* the obstacles thus giving excessively long paths as a result.

In summary of this taxonomy, we can conclude, as was already mentioned in the introduction (Fig. 1.3), by saying that to evaluate the degree of difficulty of a motion planning problem, two fundamental questions will have to be answered:

(1) What is moving, a piano or a manipulator robot? In the latter case, how many degrees of freedom does it have?

(2) Where does it move, on the plane or in three-dimensional space?

The difficulty of the problem depends on the answer to these questions. Moving from two to three dimensions implies a qualitative leap in complexity, whereas in any situation the case of the manipulator with revolute joints is intrinsically more complicated than the piano (see for example [Brooks, 1983b] for a justification of this statement). Naturally, the higher the number of the robot's degrees of freedom, the more difficult the problem will be.

Object Problem of this Application.

As a conclusion to this section, we will rigorously pose the concrete problem taken on throughout this chapter.

- The space in which the problem is solved is the ordinary three-dimensional space (3D).

- The complexity of the environment is not limited, there can exist as many obstacles as is desired.

- Robot *B* is a robot manipulator, whose number of degrees of freedom is not limited, although in the application examples

robots with six dof will be used. Revolute and prismatic joints
have been studied.

- The nature of the objects that make up the environment are
  restricted to a certain extent, since they must be objects that
  are apt to being *spherized* according to the specifications in
  chapter 2.

- The motion conditions will be the habitual ones: a start and a
  goal configuration will be given; we must find a continuous
  collision-free motion between both configurations.

- The path found will not be rigorously optimal but it will be
  acceptably short.

- The proposed method goes beyond the limits of the general
  problem (as it is been defined) since it is capable of solving
  the *adaptive motion case* (see section 4.6). In other words, it
  includes the possibility of an environment change brought
  about by the action of the robot in the course of the studied
  motion.

We have analyzed the different questions that are raised by the
general motion planning problem according to the previously
presented definition, which gives us a thorough overview of the
problem together with the possible generalizations and extensions
that have become object of study over the last few years (see section
1.2.2).

4.2 Motion Planning and Collision Detection

We have already mentioned on several occasions that the
solution to the problem of collision detection between moving objects
is a preliminary condition to cope with the robot motion planning
problem avoiding collisions in an obstacle-ridden environment. Up to
now, we have not tried to justify this statement as it upholds itself
by common sense. It is easy for anyone to understand that in order

to move without colliding into anything we must know for which placements collision occurs and for which it does not. On the other hand, in the preceding chapter, we studied collision detection deliberately avoiding —as far as was possible— references to motion planning in order to isolate collision detection and treat it as a problem in itself. Now we are going to briefly discuss the implications that collision detection has for motion planning, analyzing which characteristics we will have to demand of a method that solves the first problem if we want it to be useful in the second.

To deal with motion planning it is necessary to determine free space —if not completely, at least in part— that is, the set of free placements of the robot. In terms of a formulation that makes use of n-dimensional Configuration Space (CS), we can say that we will have to identify those configurations $(q_k)=(q_1, q_2, ..., q_n)$ that are safe, in other words, those in which intersection does not exist between the robot and the obstacles, for which it is true that:

$$B(q_k) \cap C = \varnothing.$$

Then, we will have to search among these configurations for a motion that leads from a certain initial configuration to another final one.

We can then ask ourselves how many of these free placements or safe configurations we will have to identify in order to be able to find a safe path μ. One possible answer is *all* of them, that is, CS must be completely constructed. Another possibility is to use just a certain number of them that suffices for finding a collision-free path. In any case, it will be necessary to cover a considerable portion of free space to be able to be certain that a good solution has been found.

We must keep in mind that the motion of a system made up of rigid bodies can be described as a continuous sequence of system configurations. Obviously, for a finite motion, this sequence represents an infinite ordered set of configurations. Therefore, to find a collision-free motion we must be able to ensure that each and every one of the configurations that belong to the motion are safe. In other words, we will have to solve the collision detection problem for this motion, in which we have abstracted the time parameter that describes it.

As an initial conclusion, we must flatly reject discrete methods of collision detection —like the multiple intersection test— which do not guarantee that a continuous path is safe, rather, they only *suppose* this is so from the absence of intersection in a finite collection of points belonging to the stated path. This technique may serve to achieve good visual effects in computer animation in which what interests us, more so than avoiding collisions, is that *they are not seen*. It does not matter, then, if two solids slightly collide between two images as no one will be able to see it. On the other hand, if we want to program robots at task level, it is totally inadmissible that the slightest touch occurs throughout the computed motion since that could damage the robot or its environment.

The continuous and infinite nature of spaces *FP* and *CS*, however, makes their characterization difficult. The attempts at using mathematical equations to describe obstacles in Configuration Space exactly [Donald, 1987] [Canny, 1988b] has given excessively complex algorithms as a result, with limited practical possibilities. We must, therefore, introduce approximations by employing mechanisms that guarantee that we always move in obstacle-free positions. Most of current approaches, therefore, are situated on the *safety side* in such a way that they may lose a valid solution, but they will never give an incorrect path as a solution.

In our case we will introduce the notion of *safe elemental motion* μ which will be a motion such that during its course all the generalized coordinates of the robot remain constant except a certain q_j. In other words, if the motion is given by

$$\mu: [0, 1] \to FP, \quad \text{where} \quad \mu(s) = B\,(q_1(s), q_2,(s) ..., q_n(s)), \, s \in [0, 1];$$

it is true that

$$q_i(s_1) = q_i(s_2) \quad \forall s_1, s_2 \in [0, 1], \quad i = 1, 2, ..., n; \, i \neq j.$$

Then we will say that it is *a motion in* q_j. An elemental motion for joint q_j between configurations Q^L and Q^M will be expressed as:

$$\mu[q_j] : Q^L = (q_1, q_2, ..., \alpha_j ..., q_n) \to Q^M = (q_1, q_2, ..., \beta_j, ..., q_n),$$

and it will be also called a q_j-*motion*.

Our algorithm will look for a solution path that is made up of successive safe elemental motions. Therefore, instead of deciding if a certain elemental motion is safe —a problem which we completely solved in the previous chapter— we will have to compute the limits of an elemental motion on a certain coordinate q_j that starts at a certain start configuration $(q^S_k)=(q^S_1, q^S_2, ..., q^S_n)$.

a) Case with Two Spheres

We are first going to consider the computation of those limits for the case of two spheres B and C. If the joint is prismatic, the valid placement range for the center of the sphere in motion will be a line segment in space —whose ends are given by the physical limits of the joint— from which we will have to eliminate the segment between the two points —if they exist— situated at a distance from the center of another sphere that equals the sum of their radii. In the case of a revolute joint (Fig. 4.1) we will have to exclude the arc of circumference defined by points E and F whose distance to the center of C is equal to $R_B + R_C$, and, naturally, we must also take into account the constraints introduced by the physical limitations of the links.

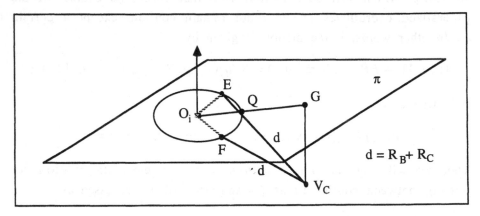

Figure 4.1 Limits in the case of a revolute joint and two spheres.

b) Case with Two Objects

In the case of two objects, we must introduce certain modifications to the collision detection algorithm described in the preceding chapter. The fundamental difference lies in the fact that we can no longer give a single answer in the sense that either there is no collision, or indeed there is, or we are still unable to ensure it. Now the output of our algorithm will be made up of three types of intervals: ones in which we can ensure that indeed there is no collision, others in which we can assure that there is a collision, and others in which we will have an indetermination at the level of representation used. We call these intervals *Free regions (F)*, *Prohibited regions (P) and Indeterminate regions (I)*, respectively. In the following discussion, we are going to ignore whether the coordinate q_j that defines the elemental motion corresponds to a prismatic joint or a revolute joint. We will refer to an interval for that q_j in Configuration Space with the clear idea that it will be equivalent to a line segment or an arc of circumference in real space, depending on the case.

Let us consider, in the first place, that we are studying collisions between two solids B and C by making use of a given exterior representation for each one of them. Solid B has a motion in q_j starting from a certain initial configuration $(q^{S_k})=(q^{S_1}, q^{S_2}, ..., q^{S_n})$, and we are going to suppose that by its physical nature, the joint permit motions in an interval $[(q^{O_k}), (q^{T_k})]$.

Each pair of exterior spheres that we may compare will give us a free region —if there is no collision throughout the entire motion— or two free regions —if it is necessary to eliminate a certain interval—. So, the final free regions for the two objects will be those that come about as a result of intersection of all free zones for each pair of spheres (Fig. 4.2). These regions will logically depend on the concrete levels of exterior representations used.

Similarly, if we use two interior representations, as a result each pair of interior spheres define a prohibited zone, which may be empty. The union of all of them will give us the prohibited zones which correspond to the two solids for the given interior representations. The zones of interval $[(q^{O_k}), (q^{T_k})]$ that have not

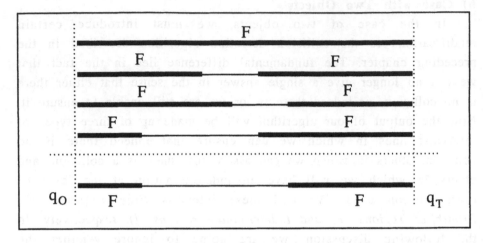

Figure 4.2   Total free region as intersection of free regions

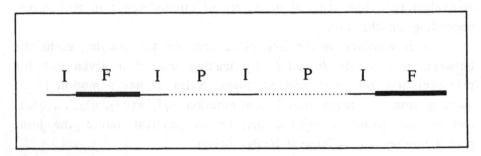

Figure 4.3   Three types of regions in an elemental motion

been defined as free or prohibited will be indeterminate (Fig. 4.3), and they will correspond to intervals in which there is collision for some exterior spheres, but there is none between any of the interior spheres.

Logically, a safe elemental motion will have to take place on the inside of one of the free intervals, since we cannot guarantee collision absence for an indeterminate interval. If the free intervals available in a given moment are not sufficient for finding a satisfactory safe path, our problem will consist in trying to clear up the uncertainty in indeterminate intervals, in order to enlarge free intervals.

The scheme for this method is very similar to the one presented in the general algorithm for collision detection. Since what we need are free intervals, we will begin by only using exterior representations which will give us free zones, on the one hand, and zones that we consider indeterminate, on the other. These representations will be progressively improved until we reach the first threshold. Note that for each new representation, we will progressively reduce the size of the considered intervals to limit ourselves only to those intervals which are, in that moment, indeterminate. Once we have confirmed that a zone is free, it will always be so regardless of the representation that we use.

Once the first threshold in the accuracy of the exterior representation is reached, we will move on to use interior spheres to compute which portions of the indeterminate zones are really prohibited zones. Once the second threshold is reached, we will proceed in a similar way to the *hard* case in the collision-detection algorithm. By means of crossed checks, we will progressively decide which representations should be improved in order to reduce the size of the indeterminate zones as quickly as possible, until we find a safe motion or we can affirm that there is not one.

c) Case with Several Objects

In our case we will have several objects which will be those that belong to the robot or to the obstacles in its environment. In the previous chapter, we have already shown that for an elemental robot motion, all of these objects can be divided into two groups: some that remain at rest during the motion, and others that all move together as if they were a single rigid body.

Then, the previous conclusions for the case of two objects can be immediately generalized for the present case. Each pair of considered objects —one in motion and another at rest— will define a set of free regions and a set of prohibited regions. Then, the free intervals that correspond to the whole elemental motion are obtained by the intersection of the free intervals for each pair of objects; whereas, the prohibited intervals of the entire motion will be the union of the prohibited intervals for the pairs of objects. Finally, the indeterminate inervals for the motion will be the ones left over, which have not been classified as free or prohibited.

To be able to adequately apply the described technique to solve the indeterminate problem, it will now be necessary to keep track of which objects are responsible for the existence of an indeterminate interval. As we know, one of these intervals results as a consequence of collision between the exterior representations of certain object pairs, whereas there is no collision for the corresponding interior representations. If we know the precise objects for which this situation takes place for a certain indeterminate interval, we will improve their representations in accordance with the previously explained scheme, while the representations of the rest of the objects remain unaffected.

To designate the *responsible* object of a certain indeterminate interval, it will be sufficient to find the intersection between this interval and the non-free intervals obtained for each pair of objects by using only exterior spheres. The objects for which this intersection is not empty will be responsible for the indetermination, since some of their exterior spheres collided —due to being in a non-free interval— and, at the same time, their interior spheres do not collide, due to being in an indeterminate interval.

4.3 Computing a Path in a *CS* Plane

Let us first hypothesize that our problem is limited to the case of a robot with only two degrees of freedom, so that any of its possible configurations would be described by means of two generalized coordinates q_i and q_j. The solution path would be contained in the two-dimensional *CS* defined by these coordinates.

a) The basic heuristic search

In order to explain the method more clearly we will follow the example shown in Fig. 4.4. Let S be the point representing the start configuration and G the point corresponding to the desired goal configuration. We wish to find a safe path joining S and G that is composed of a sequence of elemental motions, i.e., of line segments parallel to the CS coordinate axes.

First a lattice is thrown across CS, it will serve to define the configurations that are to be explored. We can structure the points on this lattice as a graph, in this way S would be the start node in the graph. From S we define a first direction of motion: along q_i or along q_j. Let us assume that we have chosen the vertical direction along q_j. Then, points A_1, A_2, ..., A_{23} will be successor nodes of S. Similarly, the successors of A_{16}, for instance, will be B_1, B_2, ..., B_{20}, and those of B_{10} will be C_1, C_2, ..., C_{16}, and so on.

Basically our approach follows the steps of a Best-First heuristic-search procedure in its specialized version called $A*$ [Pearl, 1984]. To begin with, node S is expanded, that amounts to computing the limits of the two safe elemental paths crossing S. The points on these paths that belong to the lattice are identified as successor nodes of S and they are added to the OPEN list. It must be noted that to compute the limits of any safe elemental motion, the previous collision detection algorithm must be applied. We have shown these limits as arrows in Fig. 4.4

Next the value of the evaluation function f for each node in OPEN is computed and the node for which f is minimum is selected. Let us assume it is node A_{16}. It is removed from OPEN, placed on CLOSED, and expanded. To do so, an elemental q_i-motion is computed and nodes B_1, B_2, ..., B_{20} are identified. They are put on OPEN and the value of f is found for each one of them.

The process continues basically following the previous steps, except for the introduction of some heuristic strategies that are discussed in the following paragraphs. One of these heuristic techniques allows us to refine the lattice when the possibility of a *narrow gap* between two obstacles is detected. In the example, the fact that the free intervals passing through B_7 and B_{14} are limited by two different obstacles is detected, and so the lattice resolution is locally made finer between points B_8 and B_{13}, to generate nodes B_9–B_{12}. Similarly, as the goal G may not be on the lattice as defined initially, it will be refined so that G becomes a node.

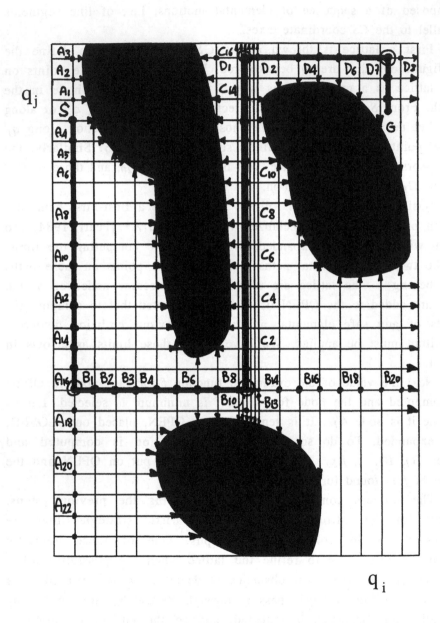

Figure 4.4   Searching for a path in a *CS* plane

b) The heuristic evaluation function

According to $A*$ algorithm the nodes defining the final solution $(A_{16}, B_{10}, C_{16}, D_9)$ have been selected because at a certain moment the value of certain *heuristic evaluation function* f for each one of these nodes was the minimum among all available partial solutions. To be correctly defined f must serve to detect the most promising potential solution. For a given node N_k it can be expressed as:

$$f(N_k) = g(N_k) + h(N_k),$$

where $g(N_k)$ is the overall cost along the path up to node N_k, and $h(N_k)$ is an estimate of the cost of the best path from N_k to the goal node. $g(N_k)$, in its turn, is an additive cost and can be expressed recursively in terms of the cost $c(N_k, N_{k-1})$ of traversing the arc from N_k to N_{k-1}:

$$g(N_k) = g(N_{k-1}) + c(N_k, N_{k-1}) \quad \text{and} \quad g(S) = 0.$$

Usually, other approaches to motion planning that make use of an $A*$ search take a *distance* between two points in CS as a measure of the cost $c(N_k, N_{k-1})$ between the corresponding nodes in the search graph, and define $h(N_k)$ as the distance between a point and the final configuration G. To our knowledge, in all previous approaches these distances were always defined as ordinary distances in CS, i.e., they are expressed in terms of joint coordinates (see for example [Lozano-Pérez, 1987]).

In our opinion this kind of measure is lacking an actual physical sense and may lead to absurd tentative paths. On the one hand, there might not be dimensional homogeneity among joint coordinates (in [Lozano-Pérez, 1987] only revolute joints are used) because revolute joints — expressed as angles— coexist with prismatic joints —expressed as distances—. It is obvious that comparing or adding an angular distance with a length does not make any sense (the question of how to unify both magnitudes is a rather obscure issue). On the other hand — now focusing on angular coordinates— a reduction in angular distances in CS may not reflect an actual approach of the robot towards its final placement in real space: just imagine a 6 d.o.f. robot that starts its solution motion by changing its last three wrist coordinates to make them closer to their final goal values, this motion would reduce the distance in CS between the robot configuration and the goal configuration G, but for an observer in real space this motion would be just a change in the orientation of the hand of the robot, while the whole robot remains far from its final position; moreover, the orientation of the wrist will very probably have to be changed again later to avoid an obstacle in the course of the complete motion.

For all these reasons, we propose as a more adequate measure in CS the use of a distance function that refers to distances in real space. This however, implies that a proximity problem should be solved [Leven and Sharir, 1987b], which results in a complex computation for the general case [Bobrow, 1989], [Gilbert and Foo, 1990] (actually this is the reason why most of previous approaches used distances in CS). Fortunately, the spherical model allows us to use distances in real space in a simple way: a valid measure of the distance between two robot configurations can be expressed as the average of the distances between the two corresponding locations for each object $Bj_i(Q)$. Therewith, the cost function c for nodes N and N' corresponding to configurations Q and Q' is given by:

$$c(N, N') = d(B(Q), B(Q')),$$

where we define

$$d(B(Q), B(Q')) = [\ \Sigma_i\ (\ \Sigma_j\ d(Bj_i(Q), Bj_i(Q')))]\ /\ \Sigma_i\ n_i,$$

with i extending over $i= 1, 2, ..., n$ and j over $j=1, 2, ..., n_i$.

To evaluate the distance between two locations of the same object Bj_i, we use the distance between the centers of their two smallest enclosing spheres (first exterior representations). For distant objects the above expression is simplified by using a unique smallest enclosing sphere for each link B_i, while for close objects more accurate representations, using more spheres, are used for each object Bj_i.

Similarly, the heuristic evaluation function h for a node N and a configuration Q might be defined as:

$$h(N) = d(B(Q), B(Q^G))$$

At this point our strategy departs from the classical A * algorithm. We pursue a finer selection of the node to be expanded in each case, otherwise, if this selection is not made very carefully, the OPEN list may contain an excessive number of elements, resulting in too many simultaneously open paths. To attain this end, when a node is expanded, its successors are not added directly to OPEN, but they are stored in an auxiliary list called the *candidate list* for the analized node. For instance, in Fig. 4.4, the candidate list for A_{16} is $(B_1, B_2, ..., B_{20})$. Then, the range of the safe elemental motion for each node in the candidate list is computed, and a *minimum distance to G* is assigned to each of these ranges. This minimum distance is calculated in the following way:

$$d_{min}(B(Q : q_k), B(Q^G)) = [\Sigma_i (\Sigma_j d_{min}(Bj_i (Q : q_k), Bj_i (Q^G)))] / \Sigma_i n_i,$$

where by $(Q : q_k)$ we mean the q_k-motion passing through Q, and therefore $B(Q : q_k)$ is the volume swept by the robot during this motion. The computation of $d_{min}(B(Q : q_k), B(Q^G))$ is done, as in the above case, by using the centers of the smallest enclosing spheres, since we know how to compute the minimum distance between a fixed sphere and a moving sphere that follows an elemental motion. Then, instead of using the previous expressions for h, the function h for a node is given by the minimum value of all d_{min} for the nodes in its candidate list; and only when a node in OPEN is selected its candidate list is added to OPEN.

The resulting strategy favors the selection of elemental motions for the first joints, since for most robots these joints govern the movement of the main body of the robot, while the last three wrist joints control the orientation of its hand.

c) Channel definition

When a solution path has been found, it is *broadened* by defining a *solution free channel* (Fig. 4.5). Note how an elemental path can be replaced by any other subpath within its corresponding

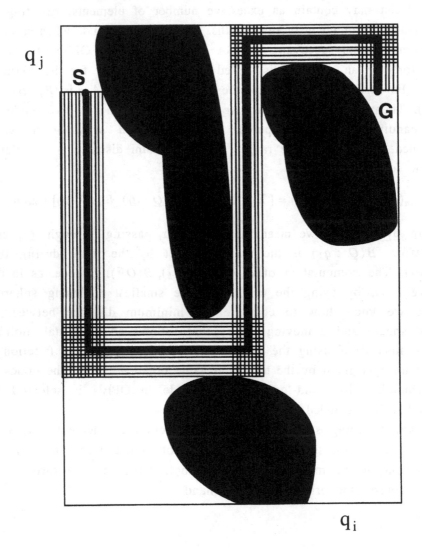

Figure 4.5   A solution free channel in CS.

channel without affecting the other elementary paths composing the complete path; in this way, our solution domain has been enlarged.

d) Terminology

We will often refer later to the subproblem of finding a safe path in a CS plane as defined by q_i and q_j, for a set of fixed values for the rest of coordinates, and between two given configurations Q^L and Q^R. This problem will be denoted as:

$$\mu[q_i/q_j]: \quad Q^L \to Q^R, \quad \text{or}$$

$$\mu[q_i/q_j]: Q^L = (q_1, ..., \alpha_i, ..., \alpha_j, ..., q_n) \to Q^R = (q_1, ..., \beta_i, ..., \beta_j, ..., q_n),$$

If one of the coordinates of the subgoal is not defined, for instance β_j, we will substitute its unknown final value by the symbol (?).

4.4 Description of the General Method

The proposed general method for robot manipulator motion planning in 3D is based on successive applications of the previous algorithm for different C-Space planes.

a) Basic strategy

For the sake of clarity in the presentation of the algorithm, we will follow an example that, without loss of generality, will allow us to better describe its different steps. We wish to find a path $\mu \subseteq FP$ for an n d.o.f. robot B connecting the start and goal configurations:

$$Q^S = (\alpha_1, \alpha_2, ..., \alpha_n) \to Q^G = (\varepsilon_1, \varepsilon_2, ..., \varepsilon_n).$$

I) First of all, the problem will be solved for a 2 d.o.f. *partial robot* (denoted as $\{B_1, B_2\}$), which is defined as composed only of the first two elements of the robot (*cutting* it through the third joint). That is, we solve the problem:

$$\mu[q_1/q_2]: \quad Q^S = (\alpha_1, \alpha_2) \to Q^G = (\varepsilon_1, \varepsilon_2); \quad \text{for partial robot } \{B_1, B_2\}.$$

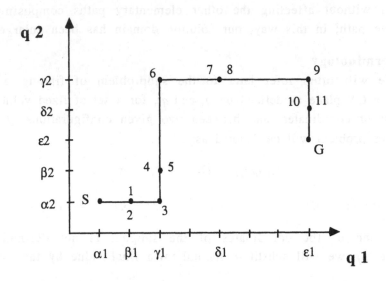

Figure 4.6   Planning in plane q_1/q_2 for partial robot $\{B_1, B_2\}$

This problem can be solved by straightforward application of the previous algorithm for a CS plane. Fig. 4.6 shows the resulting solution path which is composed of four elemental subpaths.

II) Then we consider the partial robot $\{B_1, B_2, B_3\}$, i.e., a further link and its joint are added. Starting from the initial configuration the path computed for $\{B_1, B_2\}$ is checked to find out up to what point it is still valid for $\{B_1, B_2, B_3\}$ (this is done for the initial value of the third joint $q_3 = \alpha_3$). Let us assume, for instance, that the path is still free from Q^S up to $Q^1 = (\beta_1, \alpha_2, \alpha_3)$. The same check is verified for $q_3 = \varepsilon_3$, starting at Q^G and moving backwards. Let us assume now that the path is free from Q^G back to $Q^{11} = (\varepsilon_1, \delta_2, \varepsilon_3)$.

III) Now the problem is to find some values for q_3 in such a way that the computed path for $\{B_1, B_2\}$ between Q^1 and Q^{11} is also valid for $\{B_1, B_2, B_3\}$. Each elemental subpath is analyzed and a new planning is made in one of the C-Space planes defined so far by q_3: q_1/q_3 or q_2/q_3.

To do so we first consider the q_2-motion between Q^1 and Q^3, reducing the problem to the following subproblem:

$$\mu[q_1/q_3]:\ Q^1{=}(\beta_1,\ \alpha_2,\ \alpha_3)\ \rightarrow\ Q^3{=}(\gamma_1,\ \alpha_2,\ ?).$$

As we are moving inside plane q_1/q_3 (q_2 is constant and equal to α_2) the previous algorithm can be applied to find a value $q_3{=}\beta_3$ such that the already computed subpath between Q^1 and Q^3 in q_1/q_2 is safe. In Fig. 4.7 it can be seen how this motion has been decomposed into two elemental motions on q_1/q_3:

$$\mu[q_3]:\ Q^1=(\beta_1,\ \alpha_2,\ \alpha_3)\ \rightarrow\ Q^2=(\beta_1,\ \alpha_2,\ \beta_3),$$
$$\mu[q_1]:\ Q^2=(\beta_1,\ \alpha_2,\ \beta_3)\ \rightarrow\ Q^3=(\gamma_1,\ \alpha_2,\ \beta_3),$$

IV) In a similar way, a safe path connecting Q^3 and Q^6 is found, but now, as it is an elemental q_1-motion, planning is performed on plane q_2/q_3, for $q_1{=}\gamma_1$.

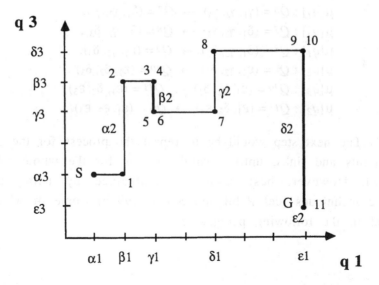

Figure 4.7   Motion in q_1/q_3 for partial robot $\{B_1,\ B_2,\ B_3\}$

In Fig. 4.8 the resulting motion can be seen, it is composed of three elemental subpaths:

$$\mu[q_2] : Q^3 = (\gamma_1, \alpha_2, \beta_3) \rightarrow Q^4 = (\gamma_1, \beta_2, \beta_3),$$
$$\mu[q_3] : Q^4 = (\gamma_1, \beta_2, \beta_3) \rightarrow Q^5 = (\gamma_1, \beta_2, \gamma_3),$$
$$\mu[q_2] : Q^5 = (\gamma_1, \beta_2, \gamma_3) \rightarrow Q^6 = (\gamma_1, \gamma_2, \gamma_3).$$

V) In Figs. 4.7-4.9 a solution is shown for $\{B_1, B_2, B_3\}$ that is composed of 12 safe elemental subpaths, they have been computed by applying the previous strategy to all the subpaths contained in the path resulting from step I. In each plane, lines of different width have been used to show a change in the value of the third coordinate not defining the plane (its value is indicated for each subpath). To summarize, these 12 safe elemental subpaths are:

$$\mu[q_1] : Q^S = (\alpha_1, \alpha_2, \alpha_3) \rightarrow Q^1 = (\beta_1, \alpha_2, \alpha_3),$$
$$\mu[q_3] : Q^1 = (\beta_1, \alpha_2, \alpha_3) \rightarrow Q^2 = (\beta_1, \alpha_2, \beta_3),$$
$$\mu[q_1] : Q^2 = (\beta_1, \alpha_2, \beta_3) \rightarrow Q^3 = (\gamma_1, \alpha_2, \beta_3),$$
$$\mu[q_2] : Q^3 = (\gamma_1, \alpha_2, \beta_3) \rightarrow Q^4 = (\gamma_1, \beta_2, \beta_3),$$
$$\mu[q_3] : Q^4 = (\gamma_1, \beta_2, \beta_3) \rightarrow Q^5 = (\gamma_1, \beta_2, \gamma_3),$$
$$\mu[q_2] : Q^5 = (\gamma_1, \beta_2, \gamma_3) \rightarrow Q^6 = (\gamma_1, \gamma_2, \gamma_3),$$
$$\mu[q_1] : Q^6 = (\gamma_1, \gamma_2, \gamma_3) \rightarrow Q^7 = (\delta_1, \gamma_2, \gamma_3),$$
$$\mu[q_3] : Q^7 = (\delta_1, \gamma_2, \gamma_3) \rightarrow Q^8 = (\delta_1, \gamma_2, \delta_3),$$
$$\mu[q_1] : Q^8 = (\delta_1, \gamma_2, \delta_3) \rightarrow Q^9 = (\varepsilon_1, \gamma_2, \delta_3),$$
$$\mu[q_2] : Q^9 = (\varepsilon_1, \gamma_2, \delta_3) \rightarrow Q^{10} = (\varepsilon_1, \delta_2, \delta_3),$$
$$\mu[q_3] : Q^{10} = (\varepsilon_1, \delta_2, \delta_3) \rightarrow Q^{11} = (\varepsilon_1, \delta_2, \varepsilon_3),$$
$$\mu[q_2] : Q^{11} = (\varepsilon_1, \delta_2, \varepsilon_3) \rightarrow Q^G = (\varepsilon_1, \varepsilon_2, \varepsilon_3).$$

VI) The next step would be to repeat the process for the rest of robot joints and links, until a valid solution for the whole robot is obtained. However, best results are obtained by using certain heuristic techniques [del Pobil and Serna, 1994b], some of which are outlined in the following paragraphs.

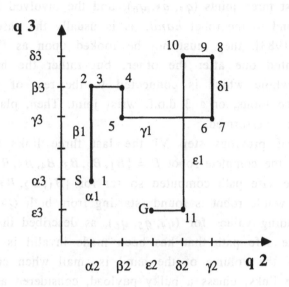

Figure 4.8 Motion in q_2/q_3 for partial robot $\{B_1, B_2, B_3\}$

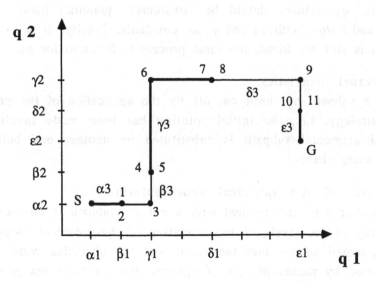

Figure 4.9 Motion in q_1/q_2 for partial robot $\{B_1, B_2, B_3\}$

b) Planning for the robot hand

If the last three joints (q_4, q_5, q_6), and the involved links (B_4, B_5, B_6), correspond to the robot *hand*, as is usually the case [Milenkovic and Huang, 1983], they must not be looked upon as three different elements jointed one after the other, but rather the hand must be taken as a whole which is connected to the rest of the robot by means of three joints, or a 3 d.o.f. wrist joint. Then, planning for the entire hand is performed.

Instead of previous step VI the last three links are added at once to make the complete robot $B = \{B_1, B_2, B_3, B_4, B_5, B_6\}$. Then, the portion of the safe path computed so far for $\{B_1, B_2, B_3\}$ that is still valid for the whole robot is found (starting from both Q^S and Q^G with the corresponding values for (q_4, q_5, q_6), as described in step II). The portion of the safe path that has been made invalid is generally not very long, as the volume of the hand is small when compared with the first three links, unless a bulky payload, considered as part of B_6, is being carried. Then, as before, motion is planned for planes in which q_4 is involved: q_1/q_4, q_2/q_4, or q_3/q_4, keeping always q_5 and q_6 as constants. In case a solution cannot be found for any of the subpaths, q_5-motions should be considered, planning inside q_1/q_5, q_2/q_5, and q_3/q_5, with q_4 and q_6 as constants. Finally, if a complete solution is still not found, the same process is followed for q_6.

c) Channel heuristics

If a subpath has been cut off by the application of the previous basic strategy, i.e., the initial solution has been made invalid, the blocked elemental subpath is substituted by another one belonging to the same channel.

d) Levels of the spherical representation

Another heuristics to deal with a cut off subpath is obviously the improving of the levels of representation of the spherical model. We have assumed so far that the robot and the obstacles were always represented by means of sets of spheres, but we have not gone into the details of how the levels of representation should be managed. Efficiency requires top level representations that use just a few spheres, while accuracy calls for low level representations using many spheres. As usual, a trade-off is necessary. In addition, starting with a too simple representation may result in a loss of many possible solutions forcing a subsequent replanning.

To start with, a common value for δ_{tot} coefficient is defined for all involved objects. This value is given by the greatest of all the first minimum values for δ_{tot}. The representation for each object having the smaller, closer value to this minimum is selected. Whenever a collision is present the representations are improved until the first threshold is reached, as it was previously described for the problem of collision detection. Naturally, only the representation for the objects involved in the collision are changed. When facing a blocked path, the interior representations come into play to determine if the path is really cut off. Similarly, they are improved until the second threshold is reached. If a conclusion still cannot be reached, the channel heuristics is used, and as a final try, cross-tests are made as in the case of collision detection.

If a robot hand can be defined, an additional heuristic is to use the smaller enclosing sphere as the first representation for the whole hand (as in Fig. 1.2, page 5). It can be refined later if needed.

e) Backtracking

In case all previous strategies fail when dealing with a cut off subpath, a backtracking process is required. That amounts to discarding the last node in the search process, returning to the immediate predecessor, and generating a new unexplored alternative solution.

4.5 Efficiency Considerations

The question of evaluating the efficiency of the present algorithm is not a trivial one. There is a trend among certain researches in motion planning —sometimes referred to as *geometric* or *algorithmic motion planning*— with the aim of finding exact theoretical algorithms for motion planning problems (some examples of these are mentioned in section 1.2.2). Though these approaches are really interesting from a theoretical standpoint — since their complexity can be expressed exactly— they generally have limited results when we try to apply them to real-life situations, if we can apply them at all.

The rest of researches in the field, which try to accomplish achievements of a more practical nature, are forced to use approximations and heuristics in one way or another, due to the inherent complexity of the problem. This is the case of the present approach. Trying to express its complexity in a formal way, by using big-O notation for instance, is difficult and of little, if any, use. Worst-case behavior is obviously of no interest, since the aim of all the employed heuristics is just to avoid dealing with this kind of situation. Instead, we will informally discuss the expected average-case performance on the basis of our experiments.

It has already been mentioned that with less than 20 spheres per object a relative error smaller than 10% can be attained. Moreover, in practical situations the level of representation for each object is selected on a common absolute error basis, so that the global representation is balanced. That amounts to using between 5 and 15 outer spheres for the largest objects, one sphere usually being enough for smaller objects. This works in 95% of usual situations, i.e., except for fine motion or delicate maneuvers. A real-life robot arm, such as the one shown in Fig. 1.1 (page 4), is divided into about 30 objects for representation purposes. Around 10 of them are rather bulky and typically need 8 spheres per object in an average situation. For the remaining 20 objects one sphere is enough (they are small details of the design). That makes 100 outer spheres for representing the whole robot (see Fig. 1.2, page 5). In an average motion only 50 of them move, which gives 5,000 tests for an elementary motion (assuming that obstacles also have a complex geometric design and need 100 spheres).

The performance of the find-path algorithm is even more difficult to assess formally. It is strongly dependent on the relative sizes of the robot links and the nature of the environment in a way that is hard to quantify. If the volumes of the first links of the robot are greater than those of the last links, as it is usually the case, the results are better when compared with robots of a more unusual nature. Similarly, for very cluttered environments, the performance slows down. Although exact surface representations have a high computational cost, they might be used instead for such environments and very fine motions

4.6 Adaptive Motion

It must also be mentioned that the present approach is especially well suited to deal with the variant of the motion planning problem known as adaptive motion [Schwartz and Sharir, 1988]. This kind of problem corresponds to a situation in which the environment changes as a consequence of the action of the robot itself. For instance, the robot may grasp an object and translate it from its initial position to a new one. Most current algorithms for the general problem cannot deal with this kind of situation; the reason is that a new mechanical system arises: one composed of the robot *plus* the object, and the problem must be reformulated anew from the start. Adaptive motion can be managed easily with the spherical model without altering the general strategy: any new object that shows up in the scene is always represented as a new set of spheres that is added to the model; its effect is therefore equivalent to simply changing the level of representation.

4.7 Application Examples

A graphical simulation system has been used to visualize the behavior of the proposed algorithm in practical situations. Fig. 1.1 (page 4) shows an actual robot manipulator. It has been modeled using a boundary representation with polyhedra. As it is a model from real life it contains hundreds of vertices, edges and faces. Using this exact surface representation would be computationally very expensive.

Fig. 1.2 (page 5) shows the set of exterior spheres used to represent the same robot in a typical collision situation. With less than 100 spheres a rather accurate representation is obtained.

Obviously, a smaller number of spheres would suffice for general situations in which obstacles are not very close. In this way, the number of computations is dramatically reduced.

Fig. 4.10 represents the Stanford Robot Arm [Widdoes, 1974], [Paul, 1981] and its kinematic structure using natural coordinates [García de Jalón, Unda and Avello, 1986].

It is a six d.o.f. robot with five revolute joints, as given by ϕ_1, ϕ_2, ϕ_3, ϕ_4 and ϕ_5, and one prismatic joint given by d_1. We will use the set of generalized coordinates $(q_1, q_2, q_3, q_4, q_5, q_6)$ corresponding to $(\phi_1, \phi_2, d_1, \phi_3, \phi_4, \phi_5)$.

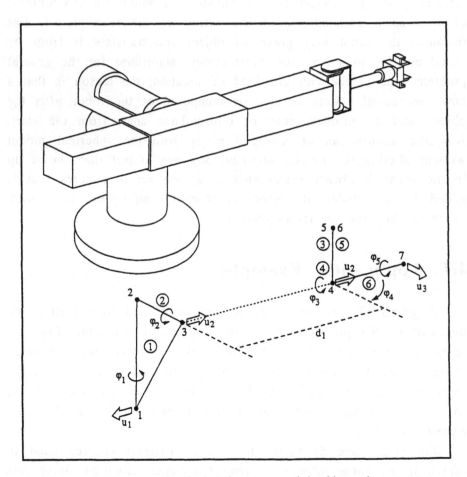

Figure 4.10 The Stanford Robot Arm and its kinematic structure

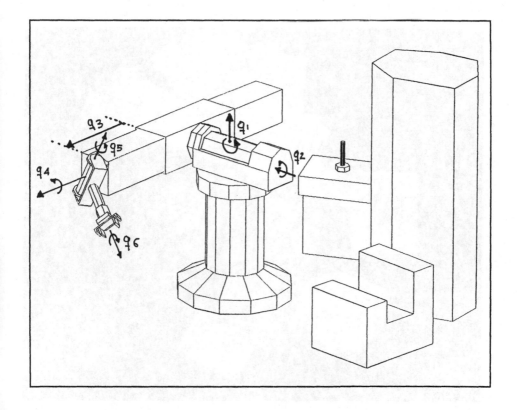

Figure 4.11   The robot in a hypothetical environment

In Fig. 4.11 the six independent generalized coordinates are shown for the robot in a hypothetical situation far away from all obstacles in its environment. For this configuration only 17 outer spheres are needed to represent all the involved objects, 10 for the robot and 7 for the obstacles, details are not necessary.

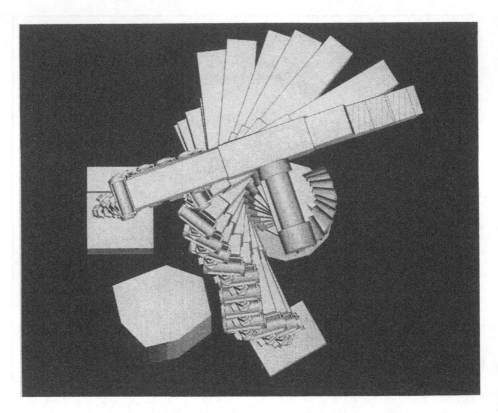

Figure 4.12   A collision-free motion resulting from the algorithm

Fig. 4.12 shows a solution collision-free motion as resulting from the proposed algorithm. This path would correspond to an initial configuration just after grasping the bolt located on the T-shaped table and a final configuration just before leaving the bolt on the U-shaped object. Fig. 4.13 shows another view for the same motion planning problem.

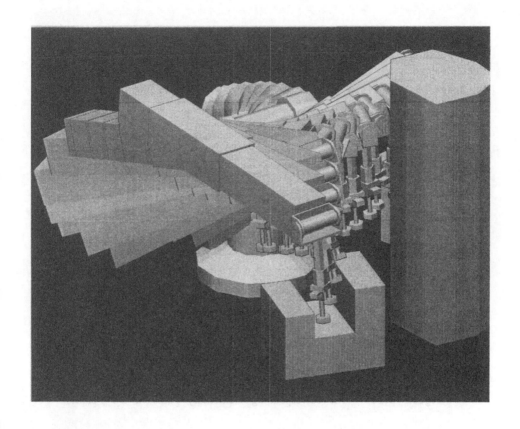

Figure 4.13   Another view for the same motion planning problem

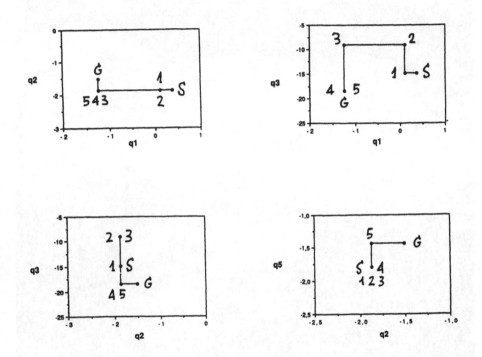

Figure 4.14   The corresponding elementary paths in CS planes

In Fig. 4.14, the corresponding elementary paths in CS are shown. Planning for four different pairs of coordinates has been necessary to find the final solution path (coordinates q_1, q_2, q_4, q_5 and q_6 are expressed in radians, while q_3 is expressed in dm). In Fig. 4.15 an instance of the underlying representation is shown for a specially difficult situation. It has to be noted that the number of involved outer spheres is just enough to determine that there is actually a path passing between the two obstacles, for this purpose the resolution of the hand representation has been refined.

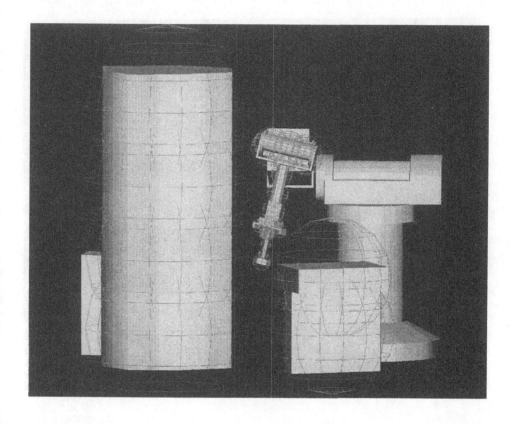

Figure 4.15   The underlying representation for a difficult situation

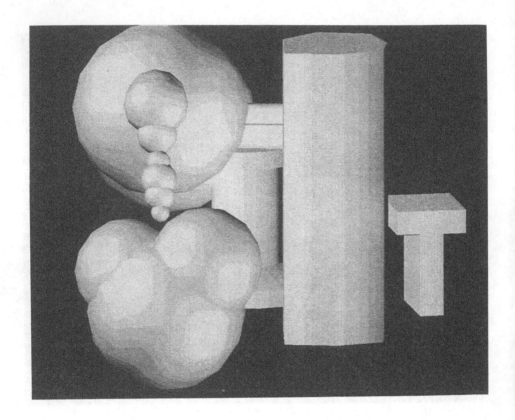

Figure 4.16   The U-shaped object is represented with 3 spheres

Figs. 4.16, 4.17 and 4.18 show details of the robot approach towards its final configuration. In Fig. 4.16 the U-shaped object is represented with only three spheres. When the robot hand carrying the bolt comes closer to the object, the representation for it is improved by using 22 spheres. In Fig. 4.18 the representation for the hand and the bolt have also been refined so that an absence of collision can be assured.

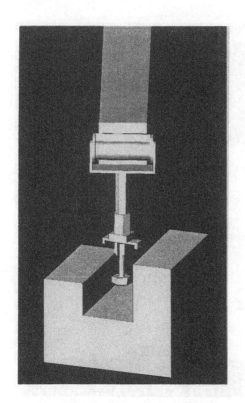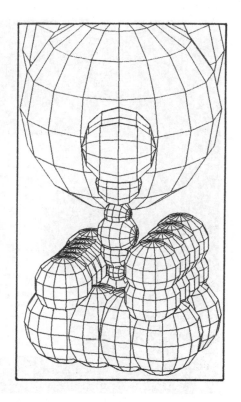

Figure 4.17   The U-shaped object is represented with 22 spheres

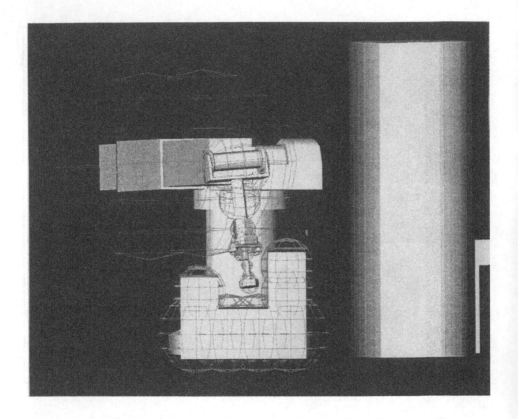

Figure 4.18   The representation for hand and bolt are refined

Finally, an application to the adaptive motion planning problem is discussed in what follows. There is an essential distinction between the situation in Fig. 4.11 and the rest: initially the bolt was just another obstacle for the robot arm; but after grasping it, it becomes part of the robot jaws. From the point of view of the spherical model both situations are basically the same: in Fig. 4.11 it was just a different representation for the hand that was used containing fewer spheres.

Chapter 5 *

Extensions to the Model

The purpose of this chapter is to introduce some extensions to the models presented in the previous chapters. The spatial representation discussed in chapter 2 had some restrictions regarding the set of objects capable of being represented. Concretely, these objects were assumed to be generated by translational sweeping of a convex polygon. The first objective of this chapter is to introduce two extensions to the spatial representation in order to obtain a more general set of physical objects that may be represented, while keeping the sphere as the only primitive.

Concretely, the extensions that are carried out seek to apply the external spherical representation to objects generated by translational sweeping of any generalized polygon. We define a generalized polygon as a closed region, convex or non-convex, limited by a set of straight-line segments and/or arcs of curves (see an example in Fig. 5.1). These extensions belong to a more ambitious project which aims at applying the spherical representation to the so-called generalized cones. These extensions make up, then, the first step of the aforementioned project and they solve problems in the 2-dimensional case.

A generalized cone consists of a planar shape, called a *cross-section,* swept along a curve, called an *axis.* A generalized cone has the following properties [Rao and Nevaitia, 1990]:

---

* Begoña Martínez Salvador is co-author of this chapter and the Appendix.

1.- The cross-section can be any arbitrary planar shape (a problem which is dealt with in this chapter).

2.- The axis may either be a straight line or a curve and it may be contained on a plane or in general 3D space.

3.- The size and shape of the cross-section can be constant or it can vary along the axis, according to a certain *cross-section function*.

4.- The angle between the section and the axis may or may not be straight.

Generalized cones have been classified according to the above-mentioned parameters [Shafer, 1983]. In this way, for example, a linear generalized cone will be one in which a linear function is applied to the section as it sweeps along the axis; a homogeneous cone has an invariant cross-section shape; and straight generalized cones will be those in which the axis is a straight line.

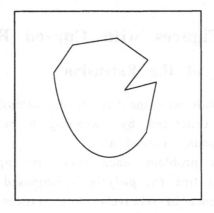

Figure 5.1   An example of non-convex generalized polygon with curved boundaries.

In sections 5.1 And 5.2 the two extensions to the external model which have been carried out are presented. First, the model is extended to polygons formed by segments and/or curves; then, the extension is carried out for non-convex polygons. These polygons may have, of course, curves and segments as boundaries.

The second objective of this chapter is dealt with in section 5.3. A generalization of the model has been developed so that it can be adapted later on to a new collision detection approach based on the four-dimensional intersection method. In this extension, the generalization of the system consists in improving an external representation using any sphere of the model. Furthermore, the role of the spherizer in this application, is to improve the representation locally in those areas where collisions have been detected. This adaptation supposes making extensions in some basic characteristics of the model originally described in chapter 2, the representations constructed will be optimal and balanced only on the local level but not on the global level, since a better representation for the whole object is not obtained; rather, the objective is to improve a region or several regions of the solid locally.

In order to solve the above problems, it has been necessary to apply some heuristic techniques as well as solve numerous problems in the field of computational geometry.

5.1. Planar Figures with Curved Boundaries

5.1.1 Definition of the Extension

This section deals with the first of the extensions carried out to represent objects generated by sweeping a polygon limited by straight segments and/or curved arcs.

To solve this problem, each curve is approximated by an enclosing polygonal line (or *polyline*) composed of line segments defined by a sequence of relatively close vertices. In this way, the spherizer system will receive a set of points as if it were a polygon made up of very small straight segments alone. The main difference is the considerably greater number of final vertices that are to be worked with. In this way, given the function or functions of one or more curves and given a number N of points for each curve, a polygon that approximates it will be built.

5.1.2 Computing the Approximation Polygon

The arcs that make up part of the boundary of the generalized polygon, are to be replaced by a series of vertices, more or less close together, which will define a polyline. Given a curve, the polygon that approximates it must fulfill both of the following requirements in order to guarantee an appropriate performance of the spherizer.

First, it is imperative for it to be an enclosing polyline of the curve. In other words, given any point P belonging to the curve, P will belong to the surface enclosed by the polyline or to its boundary. To justify this, let us remember that in the exterior representation, any point belonging to the boundary of the object must always be included in at least one of the spheres in the representation. To approximate a curve by a polyline, if we want to keep the previous property, that is, that any point of the curve belongs to a sphere, we have to ensure that the resulting polygon encloses the original curve. So, if the approximation polygon covers the original generalized polygon and the spheres cover the approximation polygon, we can guarantee that the exterior representation will cover the real rigid body.

Secondly, it is to our interest that the polyline should in a way be optimal. Since the spherizer will take the points of the polygon in order to carry out the representation, the error introduced by this approximation should be as small as possible. To compute the enclosing polyline of a curve, two cases have been distinguished: convex arcs and non-convex arcs.

When the arc is convex, the polyline is computed in the following way (Figure 5.2). Given the parametric function of the curve $(x(s),$ $y(s))$, defined on the interval $[s_0, s_f]$ and a number of points N:

1. Divide the interval $[s_0, s_f]$ into $N/2$ parts, Δs being the size of the increment.

2. Compute point P_0 of the curve for $s=s_0$ (first vertex):
$$P_0 = (x(s_0), y(s_0))$$

3. Increase s, $s = s+\Delta s$

4. Compute point P_1 of the curve for next vertex:
$$P_1 = (x(s), y(s))$$

5. Compute a middle vertex P_{01} as the intersection point of the tangents to the curve on P_0 and P_1.

6. Repeat from step 3 for the following vertices, while $s < s_f$.

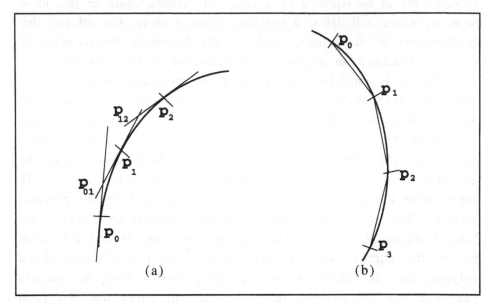

(a)   (b)

Figure 5.2   (a) Computing the approximation polyline for a convex curve. (b) Computing the approximation polyline for a non-convex curve.

In summary, for each pair of points of the curve, a third, situated between both of them, is computed as the intersection of the tangents to the curve in these points.

When the curve is not convex, the computation of the enclosing polyline is much simpler: it suffices to compute N points of the curve and to join the consecutive points to have a polygonal line (Figure 5.2).

Even though in both cases we can guarantee that the polyline will enclose the curve, we cannot yet ensure that the polygon constructed will be the smallest polygon that includes the curve. The quality of the approximation will depend on the number of vertices used to construct the polygon. Very good approximations have been obtained by using 20 or more points, though this depends on the length of the arc which is going to be approximated. The curvature of the function has not been taken into account. Taking this parameter into consideration, depending on the cases, a better approximation could have been obtained. Even though for circular arcs the use of the curvature would not be significant, it is significant for curves like the ellipse. In other words, more points will be computed where there is greater curvature. Therefore, the consideration of the curvature value to calculate the enclosing polygon will remain as a pending improvement.

Going more in depth about implementation details, when the generalized polygon is made up of two different curves, the points

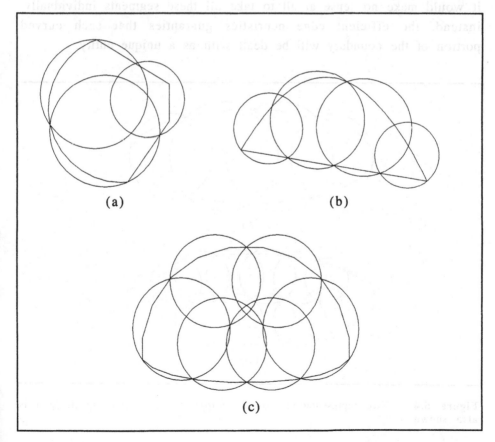

Figure 5.3   Three examples of representation for polygons with curved boundaries.

for each curve will be calculated independently, so that the sum of both is approximately N. The number of vertices that are computed for each curve will be proportional to its arc length.

5.1.3   Examples

To conclude this section, we include some results of the application of the algorithm to convex generalized polygons with curved boundaries. In Figure 5.3 three examples of exterior representations are shown. It has to be noted that the straightforward application of the method to these cases is possible thanks to the use of the efficient edge heuristics. Curved sides are approximated by a polygonal line composed of many small segments, it would make no sense at all to take all these segments individually, instead, the efficient edge heuristics guaranties that each curved portion of the boundary will be dealt with as a unique entity.

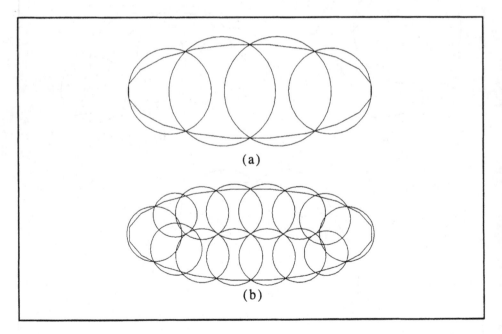

Figure 5.4   Two representations for an ellipse, the approximating polygon is also shown.

In Figures 5.3 (a) and (b), representations with only three and four circles are shown, it can be seen how the error is rather small, using only very few circles. Fig. 5.3 (c) shows the approximation polygon that has been used as input for the spherizer. Finally, Fig. 5.4 shows the polygonal approximation for an ellipse, together with two external representations using 4 and 14 circles, respectively. The polygonal approximation could be improved by defining more vertices, specially in the tips of the ellipse. Anyway, the quality of the resulting spherical representation is remarkable.

5.2 Non-Convex Generalized Polygons

In the previous section we have dealt with objects defined by swept volumes having an arbitrary convex cross-section. Now, we are going to introduce a further extension to be able to manage non-convex generalized polygons. In computational geometry, it is a general truth that working with non-convex sets complicates algorithms to a considerable degree, giving rise to the use of new techniques in order to solve numerous geometric problems (many examples that corroborate this fact can be found in the proceedings of the ACM symposia on Computational Geometry).

5.2.1 Types of Non-Convex Polygons

As was already explained in the previous section, to work with curves, a polygon that approximates the curve is obtained. In this case, concave curves will also be approximated with concave polylines. Without loss of generality, we will always speak of polygons to refer to both generalized and ordinary polygons.

Throughout this section, we will always remember that a non-convex polygon will have one or more non-convex angles that will be made up of, at least, two edges of the polygon. When we speak of non-convex angles, will be referring to any of these angles.

We are going to make a distinction between locally non-convex polygons and the general case, in which the polygon is not locally non-convex. Locally non-convex polygons are those for which only a finite number of points P exist —all of them being vertices— such that the intersection of the generalized polygon with a sufficiently small convex neighborhood of P is non-convex [Latombe, 1991]. In other words, a locally non-convex polygon can be broken down into a finite number of convex polygons. This is not possible for polygons which are not locally non-convex. In Figure 5.5 polygon (a) is locally non-convex, whereas polygon (b) is not locally non-convex.

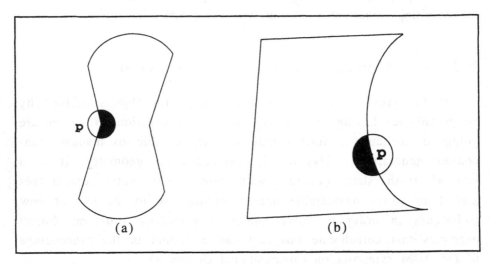

(a)   (b)

Figure 5.5 In (a) only for vertex P, the intersection of the polygon with a sufficiently small disk is non-convex. This polygons are called locally non-convex. Whereas, in figure (b), we can find infinite vertices along the curve for which the intersection of the polygon with a small neighborhood is non-convex. We say that this kind of polygons are not locally non-convex.

Although we will always work with a finite number of points, (the curves are approximated by polylines with their vertices very close together), we will be dealing with an approximation that in the limit corresponds to polygons that are not locally non-convex. The algorithms presented in the rest of this section will work for polygons without *holes*, i.e., those that are homeomorphic to a closed disk. Moreover, if the concavity itself is not convex the algorithms are not guaranteed to always work; that is, every compact subset resulting from the difference between the polygon and its convex hull, must be convex. This restriction is not satisfied, for instance, by the generalized polygon in Fig. 5.6.

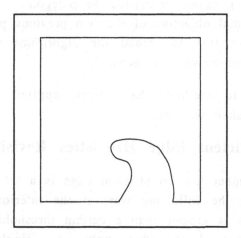

Figure 5.6   A non-convex generalized polygon with a non-convex concavity.

The spheration of concave polygons mainly poses the following questions:
- In the first place, the segments that make up the non-convex angle or angles must be detected.
- The algorithms that carry out computations to find out either the area of the object's cross-section or the error areas associated to each edge, are initially designed for convex shapes.
- Furthermore, up to now, it was assumed that the center of circles covering the edges was always located inside the polygon. We will see that when dealing with a non-convex polygon, the center is more frequently found on any of the edges or on the outside of the polygon.

When the concavity of the polygon is formed by more than two edges — this will always occur with non-locally non-convex polygons— then, we will frequently refer to all of the non-convex angles and their corresponding edges as a *concavity*.

To achieve our objective, keeping the three previous points in mind, the following are the main ideas that will be developed:

- Maintaining a data structure updated with information in order to describe each one of the non-convex angles.
- Detecting whether the center of the sphere that covers the non-convex angle is inside or outside the polygon.
- The fundamental objective of the two previous points is to have enough information to extend the algorithms so that we can work with non-convex polygons.

New heuristic methods have been applied to solve these problems, as we shall see next.

5.2.2 The Efficient Edge Heuristics Revisited

Let us remember that an efficient edge is a set of one or more edges defined in the following way: if the interior angle between consecutive edges is greater than a certain threshold angle (135° in our implementation), all the edges make up a single efficient edge, which is the same as saying that they are considered as a single edge in practice. Let us now see what happens when two of these edges belong to a non-convex angle. Obviously, the inner angle for a non-convex vertex is always greater than 180°, so it seems sensible to use the outer angle with the same threshold.

Now, if we observe Figure 5.7, the angle between the bold edges in both polygons is the same —being inner in the convex case and outer in the non-convex—; if we pay special attention to the non-convex figure, by the very definition of efficient edge, the marked edges should belong to two different efficient edges, since, as we can clearly observe, there does not exist this continuity that allows us to consider them as a single efficient edge.

However, once the first representation is obtained we can suppose that the area of greater error is that where we can find a concavity. Therefore this region must be the first to be improved since its error is greater. Nevertheless, when the spheration process was checked for different figures, with the previously explained definition of efficient edges, the results were not optimal: in most cases, the region with greatest error had not been improved.

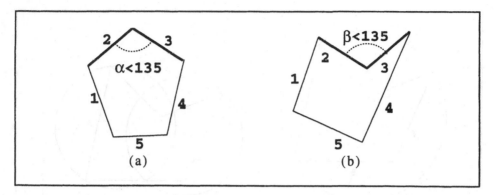

Figure 5.7   Efficient edge heuristics for convex polygons and for non-convex polygons. The same angle is a convex angle in (a) and a non-convex angle in (b). In the first case edges 2 and 3 are taken as different, while in case (b) they are part of the same efficient edge, according to the new heuristics.

In the example of Figure 5.8, the edges defining the non-convex angle give rise to one of the greatest approximation errors. Nevertheless, the spherizer did not improve this area, rather, it cut the initial list at the edges marked by arrows. The reason for not improving the area that has the greatest error can be found in the definition of efficient edges. We must remember that the algorithm chooses the worst efficient edge. Since both sides of the non-convex angle make up two edges, all of the error that is observed in this region is really divided between the two sides. Thus, both efficient edges become better than the rest of the polygon efficient edges.

After analyzing the results obtained, it turned out to be better that all of the edges that make up a non-convex angle, or a concavity, will belong to a single efficient edge. In this way, all of the error of a concavity will accumulate on this edge; as we shall see further ahead, this will also help us compute the associated error area.

In this way, going back to the examples, if we consider all the edges in the concavity as a single efficient edge, the spherizer will cut it in the following step (Fig. 5.8 (b)), and after a few steps, we observe how the representation has been notably improved in this region (Fig. 5.8 (c)).

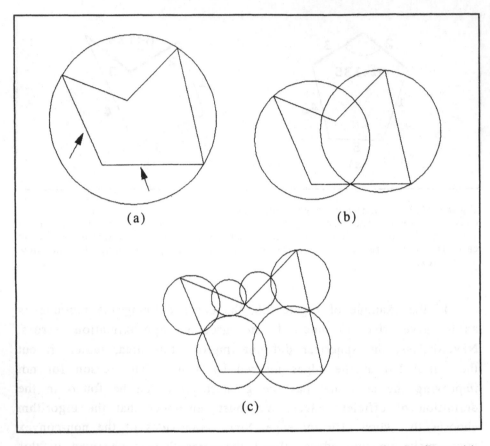

(a)   (b)

(c)

Figure 5.8   Spheration of a polygon with a non-convex angle. In figure (a), the arrows point out the two edges where the polygon would have been cut if the non-convex edges would have been considered as two different edges. Figure (b) shows how the second representation is obtained by cutting the non-convex angle. In the same way, Figure (c) shows that the smallest disks in the representation cover the region corresponding to the concavity.

Figure 5.9 is a new example of a non-convex polygon with two non-convex angles; (b) and (c) represent the second and third spheration steps, respectively. Finally, (d) shows representation number 11 for the side, where we can observe how the approximation in non-convex regions has been improved by considering each concavity as a single efficient edge.

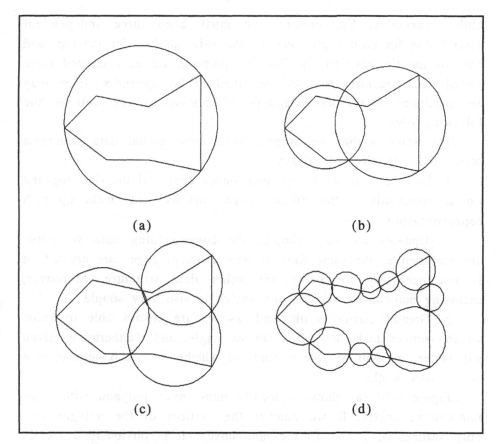

Figure 5.9   Representation of a polygon with two non-convex angles. Figures (a), (b) and (c) show the first three steps of the spherizer which obtains the second and third representation by cutting the non-convex angles.

5.2.3 Isolated Non-Convex Angles

As we already stated, we have distinguished between two kinds of non-convex polygons. In this section we are going to solve the spherical representation for locally non-convex polygons, and particularly for isolated non-convex angles. In this case, considering the three vertices that define them will suffice. In the next section we will establish the solution for the general case.

For each non-convex angle, it will be necessary to store the coordinates of the vertices that define it, indicating its order. We have chosen to describe each angle independently of already existent global variables. Furthermore, we must keep three independent descriptions for each angle: one for the side, another for the top, and one for the tip, respectively. For this purpose, we have created some global data structures that are described in the appendix, since they are pertinent to the understanding of this section as well as the following ones.

The solution that we propose with these global data structures offers the following advantages:

1. To be able to work with non-convex angles of the side, top and tip, independently of the efficient edges and lists that make up each representation.

2. Updates are very simple: the corresponding data structures are modified at the same time as new efficient edges are created. It is not necessary to access any other data structure. Moreover, accessing non-convex vertices and angles is also very simple.

3. Greater clarity is obtained, as we are always able to know which vertices make up a non-convex angle, and, furthermore, given any vertex, we will be able to find out whether or not it belongs to a non-convex angle.

Figure 5.10 (a) shows a locally non-convex polygon with three non-convex angles. If we number the vertices of the polygon (b), only vertices 3, 7 and 13 are non-convex. It is obviously a locally non-convex polygon. Note that we could easily divide it into three convex polygons (b). One possible solution would be to partition the polygon into three convex subpolygons and to spherize each one of them separately using the algorithm described in chapter 2. By using the new version of the algorithm, a remarkable result is obtained: Figures 5.10 (c) and (d) show how the spherizer, in the third representation, divides the original polygon into these three parts without using any algorithm for the partition of the polygon into convex regions.

We will analyze this example in depth in order to better understand the actions that are carried out with locally non-convex polygons. We shall start with the first representation corresponding to Figure 5.10 (a). In this initial representation, we have, as was previously stated, three non-convex angles, which will be defined in the following way: α (vertices 2, 3 and 4), β (vertices 6, 7 and 8), and γ (vertices 12, 13 and 1). Initially, the efficient edges of the unique list are made up of the following vertices:

- efficient edge A: vertices 1 and 2;
- efficient edge B: vertices 2, 3 and 4;
- efficient edge C: vertices 4 and 5;
- efficient edge D: vertices 5, 6, 7 and 8;
- efficient edge E: vertices 8, 9, 10 and 11;
- efficient edge F: vertices 11 and 12;
- efficient edge G: vertices 12, 13 and 1.

We can verify that the edges that make up a non-convex angle belong to the same efficient edge. Furthermore, we may have the case of other convex edges belonging to the same efficient edge, as occurs in efficient edge D.

After computing the error areas that are associated to each efficient edge, the system decides to cut the list at efficient edges B and G; more concretely, it decides to cut at the edges whose vertices are 2-3 and 13-1, respectively (Figure 5.10 (c)). In this way we obtain two sublists delimited by the new vertices, which we will denote as vertex 14 and vertex 15. The following vertices will make up the first sublist: 14, 3, 4, 5, 6, 7, 8, 9, 10, 11, 12, 13 and 15; vertices 15, 1, 2 and 14 will make up the second sublist.

The cut edges belong to non-convex angles. This implies that the description of these angles vary in terms of data structures in the following way: vertices 14-3-4 will make up angle α and vertex 2 will no longer belong to this angle. In fact, vertex 2 belongs to a different list altogether. The same thing occurs with angle γ, now formed by vertices 12-13-15 and, as before, vertex 1 no longer makes up part of a non-convex angle. This is the information which will be used for the correct functioning of the algorithms.

To obtain the third representation (Figure 5.10 (d)) note that the efficient edge with greatest error belongs to the first list. For this reason, edge 6-7 will be cut obtaining a new vertex. Again, this vertex, will delimit two new sublists and furthermore, vertex 6 —which defined a non-convex angle— is replaced by the new vertex in the definition of the non-convex angle.

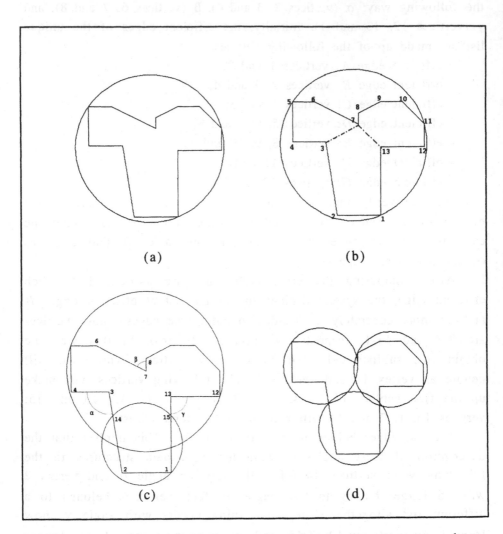

Figure 5.10   Spheration of the boundary of a locally non-convex polygon. Figure (b) shows how the polygon can be partitioned into 3 convex polygons. Figure (d) shows that the spherizer automatically divides the polygon into that three polygons without using any partitioning algorithm.

In the three cuts carried out, a non-convex vertex has ceased to define a non-convex angle while a newly created vertex takes its place. We may also have the case where the vertex that has been cut coincides with the central vertex of the non-convex angle. In this case, this angle *disappears* since the vertices that made it up now belong to different lists. In any case, these are the two possible situations that we find when working with isolated non-convex angles. We will see that in the general case there are more possible results.

To end up this example, we will insist on the fact that the three lists of representation number 3 tend to divide the polygon into three convex subpolygons, as if an algorithm for this purpose had been used.

To summarize, the basic actions to be carried out with non-convex angles would be as follows (see the Appendix where data structures are shown in greater detail):

1. In the first representation, when a non-convex angle is found, its vertices will be stored in *nsidec*. At the same time, these vertices will be marked as non-convex in *nconvex*.

2. Each time an efficient edge is cut in order to obtain a new representation, we will check whether it is made up of vertices that define non-convex angles. If the edge that is cut defines a non-convex angle, the definition of this angle is modified. In this case, the new non-convex angle is stored beginning with the first free position of *nsidec*. In *nconvex*, there will be two actions that must be carried out: on the one hand, we must delete the non-convex mark from the vertex or vertices that cease to define the angle and we must mark the new vertices as non-convex. Secondly, we update the value of *nconvex* for the vertices that belong to the angle and that now continue making up part of it since, as was already stated, the angle has been stored again in the first free positions of *nsidec*.

3. To compute the error areas associated with an efficient edge and the area of the object's base, we will also have to check whether the vertices we are working with define non-convex angles.

4. All of the previous actions are equally valid for the side, top and tip representations.

5.2.4 General Case

In this section we will analyze how algorithms get more complicated in the general case, when consecutive non-convex angles are found or the polygon is not locally non-convex. Let us remember that in this kind of polygon we have an infinite number of non-convex points, which is equivalent to speaking of non-convex curves. As we already know from previous section 5.1, we will always implement it with a finite number of points. Otherwise, we would be unable to work with them. Nevertheless, the number of points used will be great. To explain the algorithms that have been implemented, we will occasionally use non-convex curves whose number of vertices is not excessively large in order to facilitate the clarity of the explanation.

Let us assume that we are working with a concavity of n vertices, $v_1, v_2, ..., v_n$, enumerated clockwise. We will use v_k to denote the cutting point or new vertex. Now, we will state how the general case is treated when the cut to obtain the next representation takes place on the edge of vertices v_i, v_{i+1}. The concavity is divided into two parts by the new vertex. Each one of these two parts will belong to one of the two new sublists obtained by the cut:

Part 1.- i+1 vertices $(v_1, v_2, ..., v_i, v_k)$.

Part 2.- n-i+1 vertices $(v_k, v_{i+1}, v_{i+2}, ..., v_n)$.

The cutting point v_k will belong to both new concavities.

The actions to be carried out in this case are (see the definition for data structures in the Appendix):

a) Compute the number of vertices of the first new concavity.

b) Store the vertices of the first new concavity on the vector *nsidec*, *ntopc* or *ntipc* as corresponds.

c) Take into account that v_k will belong to both new concavities, therefore, we must link them as is explained in the appendix.

d) Compute the number of vertices of the second new concavity.

e) Store the vertices of these angles.

f) Update *nconvex* for v_1 to v_n.

g) Update *nconvex* for v_k according to where it was stored in *nsidec*.

Figure 5.11 shows an example of the process for the side representation in 2D when angles that are not locally non-convex exist. The polygon is made up of a combination of curves and segments. Furthermore, we have a non-convex curve as well as a convex one. Initially the figure has four efficient edges, three of them convex and one non-convex. In (a), the result of the second spheration step is shown. This result was obtained after having cut the initial polygon at both of the curved efficient edges. In this step, the non-convex concavity has been divided into two new portions. We can observe that the vertex created by the cut belongs to both concavities simultaneously. Finally, (c) shows the result after going through more steps. Each circumference corresponds to a list. If we pay attention to the non-convex curve we will observe that it is covered by four disks and, therefore, four lists which thus implies that four concavities are obtained by successively dividing the initial one.

Figure 5.12 shows another generalized polygon that is not locally non-convex. As can be observed in any of these examples, it is not possible to divide these figures into a finite number of convex polygons. In the second representation (b) one of the concavities is

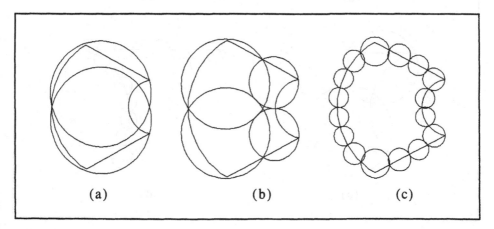

(a)   (b)   (c)

Figure 5.11   Representation of the boundary of a non-convex generalized polygon. This example combines curved boundaries with concavities.

divided into two parts. In the next representation (c), the same thing happens to the next concavity. Figure 5.12 (d) shows the result of the outer side representation after going through more steps. Finally, Figure 5.13 represents the planar outer representation of another generalized polygon for the side (a) and the top (b).

5.2.5 Computing Error Areas

Up to now we have seen what information is necessary in order to work with generalized non-convex polygons and how to update

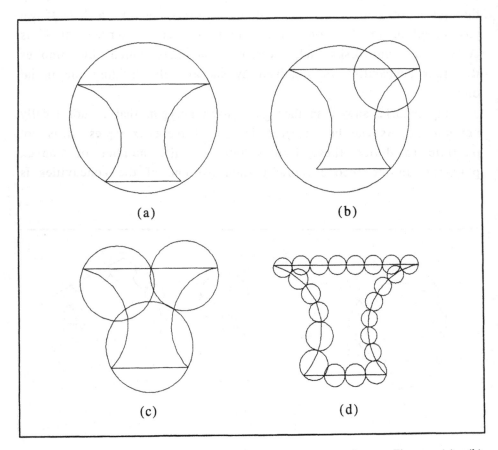

(a)   (b)

(c)   (d)

Figures 5.12   Spheration of a general non-convex polygon. Figures (a), (b) and (c) show the three first steps of the spheration process. Figure (d) shows a subsequent representation of the polygon.

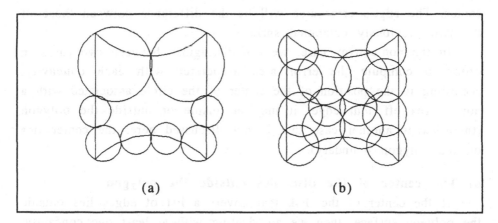

Figure 5.13 (a) Representation of the side for a general non-convex polygon. (b) Representation of the top for the same polygon.

this information as we advance in the hierarchy of representations. In this section, we will take on the problem of computing error areas. This problem becomes more complicated in the non-convex case and it is fundamental to the correct behavior of the model.

To compute error areas in chapter 2 we used the difference between the surface contained within a polygon's efficient edge and the surface of the circle sector associated with it. If we use the same algorithm in the non-convex polygon case, the incorrect computation of error areas leads to absurd cuts on efficient edges, usually being successive cuts on the same section of an edge, which gives rise to very small spheres in this area while the remaining spheres have not been improved.

In the case of convex polygons, the so-called *eff-edge-area* procedure computed error area for a convex efficient edge. This procedure, in turn, calls *edge-area* in which the error area for each particular edge of the efficient edge is computed. The efficient edge's error area will be the sum of the error area of each one of its edges.

The algorithm that implements *edge-area* works in the following way. For each edge of the efficient edge two areas are computed: the area of the circle sector defined by both of the edge's vertices, and the area of the triangle formed by this pair of vertices with the center. The edge's error area will be the difference between those of the two previously computed surfaces.

In the non-convex case, we can distinguish between two cases in order to compute the error area associated with each concavity, according to the position of the center of the circle associated with a non-convex efficient edge: it may be inside or outside the polygon (there exists another case that is not discussed here: the center lies on the polygon boundary).

a) The center of the disk lies outside the polygon

If the center of the disk that covers a list of edges lies outside the polygon surface, then we are dealing with at least one concavity. As we already now, the list can be divided into efficient-edges and, as it was said in a previous section, a concavity belongs only to one efficient-edge. Therefore, the position of the center will only influence the computation of the error area of the non-convex efficient edge, that is, the error area of the concavity. For convex efficient edges the aforementioned procedure —that is, *eff-edge-area*— will be used.

For this purpose, an algorithm —called *outside-area*— has been employed to compute the error area for non-convex cases. This algorithm proceeds in the following steps:

a) For each edge in a concavity:
   a1) Compute the area of the triangle defined by the edge and the disk center (for as many triangles as edges).
   a2) Compute the area of the circle sector defined by the two endpoints of the concavity.
b) The total error surface is the sum of all of the areas computed.

In Figure 5.14 (a) we have shaded all the error areas for the efficient edge of a polygon corresponding to a concavity. Figure 5.14 (b) represents the side representation for this polygon. It can be clearly observed how the spherical representation error in the concavity area has been remarkably reduced and, therefore, the representation has been improved.

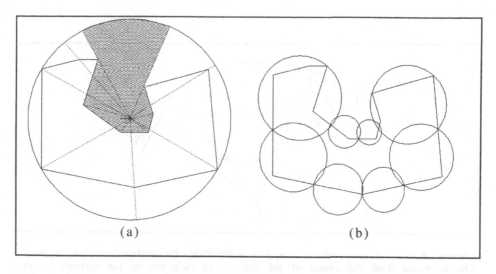

Figure 5.14 (a) Error area for the concavity of a non-convex polygon. (b) Representation of the polygon, the smallest disks cover the concavity where the big error area was located.

b) The center of the disk lies inside the polygon.

In this section we will develop a technique to calculate the error area associated with a non-convex efficient edge when the center of the circle is inside the polygon. This algorithm —called *inside_area* — is more complex since the center of a circumference can have any position in relation to a non-convex efficient edge. The input that it receives is: the number of vertices of the concavity, their coordinates and the center and radius of the corresponding circle. The idea on which its functioning is based is the break-down of the total error surface into sums and/or differences of triangle and sector areas. The problem consists in finding which points define these triangles and sectors and how to adequately combine their areas. As they are very relevant we will describe in detail the algorithms employed.

In addition to the vertices, we are also going to look for other points of the concavity edges which will delimit the surfaces we are looking for. These points are the intersections between the concavity edges and the straight lines that join the center of the covering disk with each one of the concavity vertices. Figure 5.15 shows a concavity with vertices u, v, w, x, y. The straight lines that join the

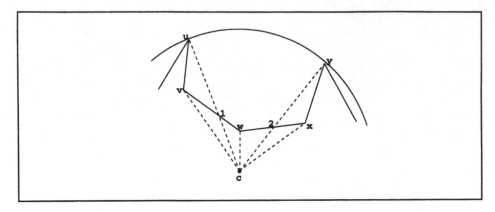

Figure 5.15   A non-convex concavity defined by the vertices u, v, w, x and y. The segments from the center of the disk, c, to each one of the vertices cut the edges in points 1 and 2.

center with each one of the concavity vertices give rise to two intersection points numbered *1* and *2*.

The algorithm will always need to know whether the point which it is working with is an original polygon vertex or a point obtained from intersection. To find this out, we will use a matrix with four rows. We will have all the points, concavity vertices and intersection points alike, stored in the first row. The second row will contain what we have called *associated points*, which are in relation with each component of the row above it, so that if the line from a concavity vertex b_i cuts one of the edges of the concavity on c_j, we will say that b_i has c_j as *associated point* and vice-versa. In the third row we will store number 1 if the corresponding point in the same position in the first row is an original concavity vertex, or 0 if it is one derived from cutting. Finally, in the last row 1 or 0 will be stored according to whether the corresponding point in the second row refers to a vertex or a intersection point.

We will initialize the matrix with concavity vertices in the two first rows, with 1 in each one of the components of the third and fourth rows (initially, each vertex has itself as associated point). With this, the initial values of the matrix for Figure 5.15 are:

u	v	w	x	y
u	v	w	x	y
1	1	1	1	1
1	1	1	1	1

Once we have initialized this structure, the algorithm is the following:

1.- For each vertex b_i: compute the intersection, if there is such, between the concavity edges and the straight line defined by the disk center —$cent$— and b_i.

2.- For each intersection point c_j found:
  • its coordinates are added in the first row of the matrix,
  • in the second row, the coordinates of the its associated point are stored (i.e., the vertex that brought about the intersection);
  • the position of this vertex is searched out in the first row of the matrix and in the very same position of the second row the intersection point will be placed (as this is the associated point for the vertex)
  • complete the third and fourth rows as corresponds.

3.- If a vertex gives rise to no intersection, this same vertex is stored in the second row (its associated vertex is itself, so to speak).

4.- Arrange all points in clockwise order. (The four rows will have to be ordered).

There exists the possibility that one vertex gives rise to two intersection points. In this case, the vertex will be repeated in the data matrix.

Once we have all the points placed in order with their associated points and flags —indicating whether they are original vertices or not— we proceed as follows:

1.- Initialize the variable *lastp* to the column index for the last point in the first row of the matrix.

2.- Initialize the variable *flag* to *false*.

3.- For each pair of consecutive points from the first row of the matrix, v_i and v_{i+1}, while $i<lastp$

3.1.- Their associated points are read (second row), p_i and p_{i+1}.

3.2.- Find whether v_{i+1} has already appeared as an associated point, and if so, check whether v_i was consecutive. If both conditions are fulfilled, we will go back to step 3 and read the next pair of points.

3.3.- If both points, v_i or v_{i+1}, are vertices, then:

    3.3.1 If both associated points are also vertices:
- call procedure *edge_area*,
- *flag=true*

    3.3.2 If any one of the associated points, p_i, p_{i+1}, is an intersection point:
- compute the area of the triangle defined by v_i, v_{i+1} and the center,
- compute the area of the triangle defined by p_i, p_{i+1} and the center,
- subtract both areas (subtraction order will be determined by *flag*).

3.4.- If any one of the points, v_i or v_{i+1}, is an intersection point, then:

    3.4.1 If both points are the same (this will occur when a vertex has two cutting points associated with it):
- call *edge_area*,
- *flag= true*

    3.4.2 If they are not the same, then:
- If p_i is located before v_{i+1} in the first row of the matrix and p_{i+1} comes after v_i in the same row, then the area is computed by *edge_area* and *flag* becomes *true*.
- If the previous condition is not fulfilled:
  - compute the area of the triangle defined by v_i, v_{i+1} and the center,
  - compute the area of the triangle defined by p_i, p_{i+1} and the center,

- subtract both areas (subtraction order will be determinate by $flag$).

3.5.- If v_{i+1} or p_{i+1} are equal to v_{lastp}, then $lastp = lastp - 1$.

4.- Finally, the areas obtained for each edge are added.

To make the algorithm clearer, next we will apply it to some examples. The first example corresponds to Figure 5.16. It represents a concavity with six vertices, b_1, ..., b_6. When we join the disk center to each one of the vertices, the intersection points c_1, c_2 and c_3 are found. Initially, we stored all of the initial vertices and intersection points in the matrix:

b_1	b_2	b_3	b_4	b_5	b_6	c_1	c_2	c_3
c_1	b_2	b_3	c_2	b_5	c_3	b_1	b_4	b_6
1	1	1	1	1	1	0	0	0
0	1	1	0	1	0	1	1	1

In relating the matrix data, we can say that vertex b_1 has c_1 as its associated point; in the same way, c_1 has b_1 as its associated point. As the line that joins the center to vertex b_2 does not intersect any of the edges, then this vertex has itself as its associated point.

As we have explained, we also need to know which of the points are original vertices and which are obtained by intercepting. This data will come from the third and fourth rows of the data matrix. Its meaning will be the following: the first six components of the matrix are original vertices, while the last three are intersection points. In other words, in position $i=1$ we have an original vertex with an associated intersection point, while in position $i=2$ there is a vertex which produces no points of intersection.

The next step would be to place the four rows of the matrix in clockwise order. The ordered matrix will be the following:

b_1	b_2	c_1	b_3	c_3	b_4	b_5	c_2	b_6
c_1	b_2	b_1	b_3	b_6	c_2	b_5	b_4	c_3
1	1	0	1	0	1	1	0	1
0	1	1	1	1	0	1	1	0

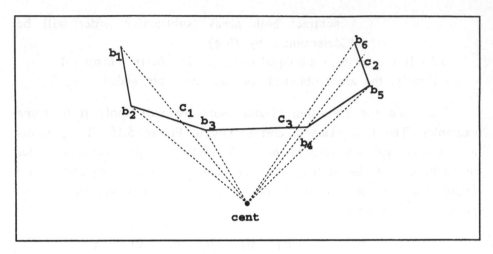

Figure 5.16   A non-convex concavity with vertices b_1, b_2, b_3, b_4, b_5 and b_6, whose associated disk has the center in point called $cent$. c_1, c_2 and c_3 are the intersection points of the line segments that join the center with each vertex.

Now we can apply the algorithm. First, initialize $lastp$ to 9 and $flag$ to $false$. Take b_1 and b_2 and their associated points c_1 and b_2, respectively. By looking up the matrix values, we learn that there is an intersection point among the four points, namely c_1. The area we are looking for will be the difference between the area corresponding to the triangle defined by the center and vertices b_1, b_2 (read from the first row of the matrix) minus the area corresponding to the triangle defined by the center and vertices c_1, b_2 (read from the second row of the matrix).

We will check if b_2 is equal to b_6 (since $lastp$ =9). If it is not, we will continue by reading the next point. Now we take b_2 and c_1 and their associated points b_2 and b_1. Once again we have an intersection point among the points. Before computing anything, we verify that c_1 has already appeared (in the above iteration) and furthermore, that it is stored before b_2 in the second row of the matrix, in which case no area for that edge will be computed and we will go on to the next iteration. If we observe Figure 5.17, the area calculated for the first edge already includes this second edge, therefore, if makes no sense to compute an area here.

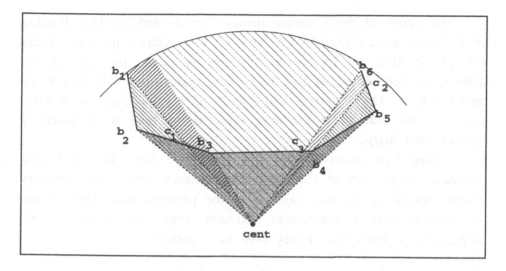

Figure 5.17   Error zones of the concavity.

In the next iteration we have points c_1 and b_3 with their associated points b_1 and b_3, respectively. Again, we have an intersection point among them. Vertex b_1 has already appeared in the first row of the matrix and it comes before point b_3, and b_3 is stored after c_1 in the first row of the matrix. The area in this case will be computed by *edge_area*, that is, as the difference between a sector and the triangle defined by vertices c_1, b_3 and the center. The variable *flag* will change to *true*.

In the following iteration, for edge b_3-c_3, we have the same situation as in the previous iteration (b_3 is stored before c_3 in the first row of the matrix and b_6 has not still appeared). Therefore, once again there is a call to *edge_area*. In this case, b_6 is equal to the element at which *lastp* points, then *lastp* becomes 8.

The next boundary portion corresponds to points c_3 and b_4 with associated points b_6 and c_2. After making all pertinent checks, area will be computed as the difference of the areas of the two triangles: the one defined by the points from the second row of the matrix with the center, minus the one defined by the points from the first row with the center (subtraction depends on *flag*). Like before, *lastp* must be updated since it coincides with c_2, now its value will become 7.

The points of the following iteration are b_4 and b_5. The situation is the same as the previous one and we will subtract the area of the triangle defined by the center, c_2 and b_5 minus the area of the triangle of vertices b_4, b_5 and the center. Again, *lastp* will have to be updated to 6. The algorithm will stop at this point since the column index in the matrix for the next value to be read is 7, which is greater than *lastp*.

Figure 5.18 shows another example where the center is totally displaced to the left of the non-convex efficient edge. The algorithm works exactly in the same way as in the previous case. The number of vertices that a non-convex efficient edge has, is not at all important, we know that it may even be a curve.

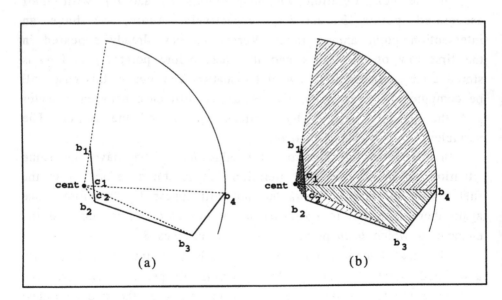

Figure 5.18 An example of error areas for a concavity, where the center of the associated disk is located on the left of the efficient-edge.

5.2.6 Examples

The last figures of this section (5.19-5.22) show 2D representations for generalized polygons defined by combinations of curves and edges, including non-convex angles. It can be seen that in

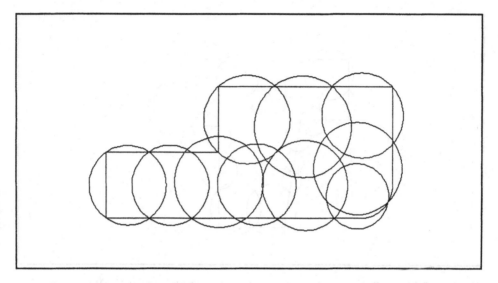

Figure 5.19 Representation for the whole area (top) of a generalized polygon. In this figure and the following one, the polygon may be divided into two convex polygons, anyway we can spherize it without partitioning the polygon.

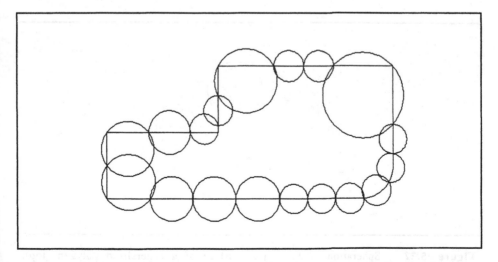

Figure 5.20 Covering for the boundary (side) of a generalized polygon.

all cases, the algorithm is very efficient as it attains a small error with a low number of disks. Some cases correspond to the covering of the polygon's boundary and others to the covering of the entire polygon surface.

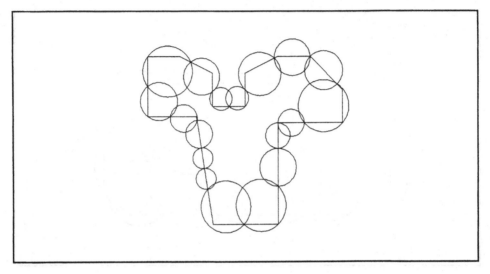

Figure 5.21 Representation of the boundary (side) of a locally non-convex polygon without making a partition into convex parts.

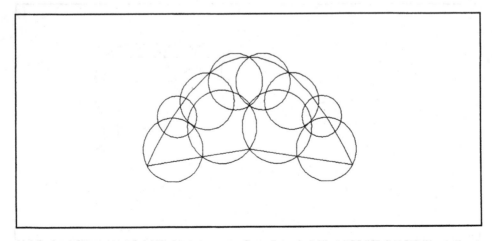

Figure 5.22 Spheration of the whole surface of a generalized polygon (top).

5.3   Application-Dependent   Local   Quality Improvement

In this section we will deal with adapting the spherical representation so as to obtain a local quality improvement. The application is based on the system's capacity to automatically obtain new representations with the desired level of approximation. In addition to making a balanced global improvement, in this new operation mode, the spherizer may receive a set of spheres whose approximation it must improve obtaining a new representation that improves the quality in those regions indicated by the concrete application.

There is a fundamental difference between this new mode —which aims at a local improvement of the representation— and the original approach. The expert spherizer could only choose the sphere which was to be improved according to certain quality criteria, in such a way that the worst zone was always improved. Now, the spherizer can choose between two operation modes. In the new mode, the sphere or spheres that it must improve are already determined by the client application. This implies that it will not choose the representation of the object (side, top, tip) to be improved either. Therefore, it will only decide the action to be carried out in order to improve a particular set of spheres.

In this way, if we use the representation together with a client collision-detection system, we will only improve those spheres that produce collision. Accordingly, the efficiency of the collision detector is improved by keeping the number of spheres very low. Starting with a representation r_i, the spherizer system will improve those spheres that produce collisions, obtaining a new representation of the object, r_j, not necessarily consecutive to r_i. In other words, between r_i and r_j there may be some intermediate representations, since between them the spherizer may have carried out more than one action.

We can say that obtaining successive representations of an object with the original mode of the expert spherizer is linear, that is, starting with representation r_i, there only exists one path to reach representation r_k, where going from one node to another means improving the global quality of the previous representation (Fig. 5.23 (a)). Now, the representation generation scheme will correspond to a tree (Fig. 5.23 (b)), where beginning with representation r_i we can choose different paths according to the set of spheres we have to improve and, thus, which action to apply. Each node from the tree is a representation which comes about after having chosen a particular action upon a set of spheres.

We do not always start with the last representation obtained in order to get a new representation of an object. This will depend on the region that we want to improve. On the other hand, beginning with the same representation, we can get different representations depending on the spheres that are substituted.

If we can obtain different offspring representations from representation r_i — a node of the tree— depending on the improved circle, and at different moments, then the representation numbers of the offspring nodes have no relationship to the parent representation number. In a call to the spherizer, the system can go to any one of the representations (any node in the tree), moving along the tree, and generate new spheres from it by substituting other spheres. Therefore, the spherizer will have to keep in memory the different representations that have been created up to now and it will have to be able to go to any of these representations with each call. That is, the present representation will no longer be the last one necessarily.

We can summarize the behavior of the expert spherizer in the new mode in the following points:

a) The system does not choose which sphere or spheres it improves by means of global quality parameters.

b) Each time the spherizer is called, one or several actions can be carried out on the same object.

c) A new representation can be generated from any of the previous representations in the tree.

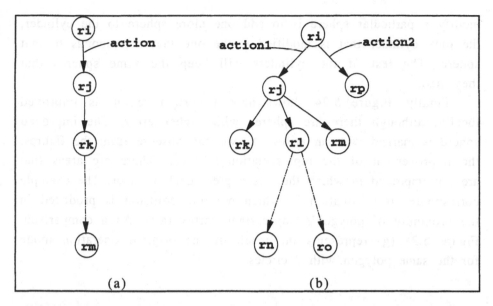

Figure 5.23   (a) Shows the linear structure resulting from the original operation mode to obtain the representations. (b) Shows the tree generated with the new operation mode, starting in the same representation, r_i, we can obtain different new representations.

We have already cited the consequences of this operation mode: local improvement of the desired region, the total number of spheres is kept low and representation generation is not linear.

In a call to the spherizer, it receives —for each object— the spheres of a given representation that it should substitute. The first step when it receives the spheres is to classify them, that is, to group them first at all into side, top and tip and secondly by cylinders (spheres brought about by the same base circle). The aim of this classification is that only one action will be carried out in order to improve a cylinder, when several spheres from the same cylinder must be substituted, since one cylinder corresponds to a unique generator circle. Naturally, all of the cylinder's spheres will be modified in this action. The second step, after grouping or classifying, is to identify the circle corresponding to each cylinder that has to be substituted.

A new feature is that the system allows different cylinders to have a different number of spheres. Therefore, if the action chosen to modify a particular sphere is to add one more sphere to the cylinder, the only cylinder that is modified is the one that corresponds to that sphere. The rest of the cylinders will keep the same spheres that they had.

Finally, Figure 5.24 shows how a certain region is improved locally, although there are spheres with higher error. The improved sphere is marked with an arrow. As we can observe from the figures, the improvement of the representation is local. There are areas that are not improved in which there is a great deal of error. The example corresponds to a situation in which possible collision is produced in the proximity of polygon's upper right vertex (a-f). As a comparison, Figure 5.24 (g) represents the result of the original operation mode for the same polygon with 7 circles.

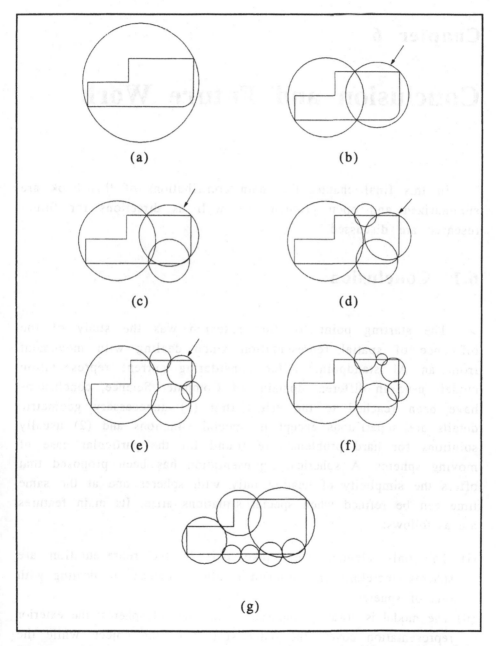

Figure 5.24 (a), (b), (c), (d), (e) and (f) show the local improvement of the representation of a polygon. Figure (g) shows a representation of the same polygon using the original mode.

Chapter 6

Conclusion and Future Work

In this final chapter the main contributions of this book are summarized and open problems as well as directions for future research are discussed.

6.1 Conclusion

The starting point for this research was the study of the influence of spatial representation when dealing with movement from an AI standpoint. After considering current representation models used in different domains of Computer Science, conclusions have been reached to the effect that (1) unnecessary geometric details are superfluous except in special situations and (2) usually solutions for hard problems are found for the particular case of moving spheres. A spherical representation has been proposed that offers the simplicity of dealing only with spheres and at the same time can be refined when special situations arise. Its main features are as follows:

(i) The only elements that take part in the representation are spheres; therefore, any problem is always reduced to dealing with sets of spheres.

(ii) The model is always composed of two sets of spheres: the exterior representation covers the outer surface of the object, while the spheres in the the interior representation are contained in the object.

(iii) A hierarchy is defined for both representations in the sense that it ranks from simple and rough to complex and accurate representations: the approximations for the object can be made better and better by using more and more spheres.

(iv) Each representation is balanced, i.e., its error zones are not concentrated at specific locations. In addition, the representation is reasonably optimal, considering that it is generated by a heuristic algorithm.

To find an optimal representation with some of these features is an NP-hard problem in Computational Geometry, even for the simpler planar case. An original set of algorithms has been developed to define the model with generality and quality, for two and three dimensions, and for inner and outer representations. In this way algorithms have been presented for covering a polygon or its boundary with circles, for packing a polygon with circles and for packing and covering solid bodies with spheres. Some results from the mathematical theory of packing and covering have been taken and extended both for two and three dimensions. A heuristic rule-based system has been designed to deal with the decision process involved in the construction of the representation. A set of parameters has been introduced that provide us with a global and local measure of the degree of accuracy of the model. It has been shown how the sequence of representations in the hierarchy converges to a zero-error limit.

This model differs from any previous spherical model, since it combines the notion of hierarchy of detail with the utilization of the sphere as the only primitive in the whole model. Though it may be argued that for representations with many spheres a typical model would be better, the representation must be defined to be complete in itself: there must be no limit to its accuracy without being compelled to appeal to other primitive objects.

The double hierarchical representation has proved to be a useful tool to simplify the solution of many important questions relevant to the motion planning problem. To show its capabilities, a new approach to planning collision-free motions for general 6 d.o.f. manipulators has been presented. The algorithm takes advantage of the simplicity of dealing only with spheres but, at the same time, due to the power of the notion of hierarchy of detail, the involved approximations are tuned up to the required accuracy in each particular case.

Particularly, the complexity of the general collision detection problem is reduced, being especially simple for the case of a manipulator. It has been shown how, in the case of moving only spheres, the swept volume intersection test becomes much simpler as it can be reduced to computing distances between fixed points and simple curves. Computing the volume swept by more complex objects is generally a much more difficult task. By making use of the uppermost simpler representations in the hierarchy, unless a refined lower level one is needed, the number of computations is dramatically reduced, moreover all the details —faces, edges and vertices— of the objects are ignored, except for the objects being close enough. This was not the case for many previous approaches. The twofold nature of the model allows a faster determination of cases of actual collision without having to refine the involved representations, keeping in this way the number of involved spheres as small as possible.

By using a modified version of the $A*$ algorithm realistic collision-free paths have been efficiently found on CS planes. The spherical model makes it possible for the utilization of a heuristic evaluation function with a real physical sense. Moreover, Configuration Space is not completely constructed, and computational cost is reduced to the strictly necessary by selecting the most adequate level of representation. A general strategy has been defined for general 6 d.o.f. robots with good results for actual robot models with complex design structures. Several heuristics have been introduced to speed up the performance of the system. The spatial representation is also especially well suited to deal with the variant of the general movers´ problem known as adaptive motion. Finally, some application examples have been presented.

6.2 Future Research and Open Problems

Some research directions aiming at either improving the present results or further applying it to new problems are suggested in what follows. The authors are already working on some of them.

The spatial representation itself could be improved by developing new algorithms to enlarge the range of objects that can be represented (chapter 5 includes a first step in this direction).. Similarly, considering more general elemental motions would be very interesting, though dealing with more general curves will lead to nontrivial problems in differential geometry.

Another interesting possibility is applying the representation —perhaps in a modified version— to detect collisions by using a four-dimensional intersection test in space-time.

Dealing with variations of the classical motion planning problem is another open possibility. The cases of coordinated motion or moving obstacles seem to be the more promising for applications of the proposed approach.

Finally, as the representation allows easy computations of interactions for objects in motion, distances, positions, etc., another future application of the spherical model is spatial reasoning. This problem may be defined as the development of techniques to manipulate descriptions of physical objects taking into account their mutual relationships, as well as their shapes, sizes, positions and motions. The hierarchy of detail defined in our model permits us to characterize a distant object by means of a few parameters: its size is given by the diameter of its enclosing sphere (first exterior representation) and its position is given by the coordinates of the center of the same sphere. When an object is closer, its different constituent parts are represented in the same way by different sets of spheres. The relationship among these sets serves to derive a description of the shape of the object, and similarly, a symbolic description of the relationship among objects can be obtained from the relationship among spheres. Some discussions in this respect can be found in [del Pobil, Escrig, Jaén, 1993, 1994].

6.3 A Concluding Reflection

To conclude, we will make a final reflection on the long-term aim of research in Intelligent Robotics. A related longer discussion —that is worth reading— can be found in the *Introduction to Robotics* by P.J. McKerrow [1991] (pp. 14-23) which, in addition, is an excellent textbook. The work described in our book —together with the work of many other researches in this field— is a contribution towards more autonomous and useful robots. We could even say more *intelligent* in the same way we say an ape is more intelligent than an insect. We may also say that in the future robots will become more and more similar to persons, if by similar we understand that they will be able to mimic many capabilities of human beings. However, science will not be able to go further, since there will always be an *ontological discontinuity* between human beings and the rest of living creatures, in the sense E. F. Schumacher discusses it in his posthumous essay *A Guide for the Perplexed* (1977). This gap in the level of being is summarized by Schumacher's expression for human beings:

$$m + x + y + z,$$

the z will always be missing for a robot. This z represents personality, fortitude, fidelity, love, the appreciation of beauty, hope, the ability for personal relationships, faith, dignity, prudence, justice, magnanimity, solidarity, ... What is this z? How can it be defined? Answering these questions would take us too far now. Let us only remind that we are not complex machines, and we are not naked apes either.

Appendix

Data Structures for Non-Convex Polygons

Due to its relevance, we will include in this appendix a description of the data structures used to implement the algorithms that allow for the spherical representation of non-convex polygons. These structures are independent of the ones used for the representations of whole objects.

In the following, we will show in detail, by using an example, which structures have been created and with what purpose. Let us suppose that we are at the beginning of the spheration process of an object. Starting with the base vertices, we identify its edges and at the same time, the concavities it may have. In the example of Fig. A.1, the base will have a non-convex angle made up of vertices 1, 2 and 3. We will have an initial matrix, called *nsidec*, which will store the concavities for the side in the following way:

1. In row 1 first store the number of vertices that make up the concavity and next, the order number of those vertices on the angle, which in this case coincide with the numeration of vertices, 1, 2, 3.

2. In the second row, we will keep the pointer to each angle vertex, which will also be 1, 2 and 3 for this example.

A second matrix, *nconvex*, will have a component (column) for each vertex and one row for the side, another for the top and a final one for the tip. Continuing with the same example, it will keep (always for the side, for the moment) 2 in the component corresponding to vertex 1, first column and first row. This means that

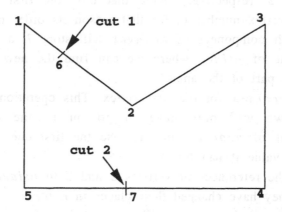

Figure   A.1

this vertex belongs to a concavity for the side and that this vertex is stored in the second component of *nsidec*. In the same way, it will have 3 and 4 in the second and third component. For vertices 4 and 5 it will have 0, indicating that none of them belong to a concavity (see Figure A.2).

Thus, given any vertex of the base, we can find out if this vertex defines a convex or non-convex angle by looking up its corresponding component in *nconvex*, that is, if it makes up part of a concavity. If the vertex defines a non-convex angle, by reading this value, we would have access to the information of the concavity which it makes up part of, whether for the side, top or tip. Rows 2 and 3 of *nconvex* have the value 0 for the five base vertices, since, in our example, neither the top nor the tip have been defined.

Continuing with the spheration process for the same example (Fig. A.1), when the first two cuts are produced, we will discover that vertex 6 defines a new non-convex angle and vertex 1 no longer defines it. In this case, the modifications to be made are the following:

1. Store a new concavity in *nsidec*, as before, first the number of the vertices, 3, and then the order of the vertices 1, 2, 3; and in

the second row the pointers to the coordinates of those vertices: 6, 2 and 3, respectively. Note that now the first vertex of the angle is vertex number 6, for that reason its order number is 1.

2. The sixth component of *nconvex* will store 7, a pointer to the component of *nsidec* where we can find the new vertex which makes up part of the angle.

3. Update *nconvex* for the first vertex. This operation will be very simple: we will only have to go on to the corresponding component of *nconvex*, in this case the first one, and substitute whatever value it had for 0.

4. Update the references of vertices 2 and 3 to *nsidec* in *nconvex*, as now they have changed their places in *nsidec*.

5. For the second cut, we will simply place 0 in the *nconvex* component for vertex 7, since it does not belong to any non-convex angle.

After all of those actions, Figure A.2 will have been modified until Figure A.3 is obtained.

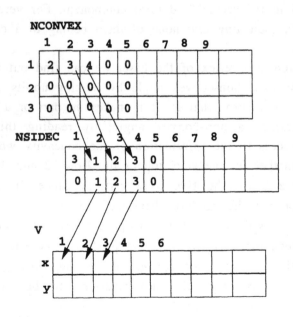

Figure A.2

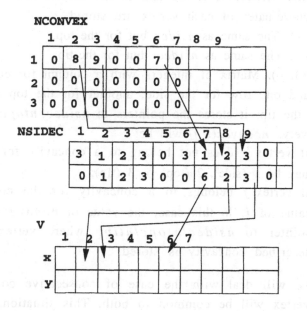

Figure A.3

Although it does not appear in the figures, we have already said that independent structures for the top and the tip are needed. Therefore, in addition to *nsidec*, we will have *ntopc* and *ntipc*, for the top and the tip respectively, defined exactly in the same way. Their function will be the same as *nsidec*. Therefore, when the first representation is carried out for the top, we will have a new data structure that is the same as the one shown in Figure A.2, but this time using *ntopc*. From now on, the explanation will be done for the side, understanding that they will be equally valid for the top and the tip. The order of the vertices has been explicitly maintained for greater clarity. It will be important to find out which vertex of the concavity we are dealing with.

Below, we gather together all of the data structures used for concavities.

- *nsidec*(2,—). Matrix that describes the concavities for the side representations. In the first row the position occupied by each one of the vertices is stored and in the second row, the pointers to the coordinates of each vertex are stored.

- $ntopc(2,—)$. The same as $nsidec$ but for the top.
- $ntipc(2,—)$. The same as $nsidec$ but for the tip.
- $nconvex(3,—)$. Matrix of integers with a column for each possible vertex and one row for the side, another for the top and another one for the tip; it stores the pointers to $nsidec$, $ntopc$ and $ntipc$, respectively. $nconvex(i, j)$ will be:

  = 0, if vertex j does not belong to a concavity for i, where i may be 1 (side), 2 (top) or 3 (tip);

  $\neq 0$, if vertex j belongs to a concavity for the corresponding value of i. In this case, the value of $nconvex(i, j)$ is the pointer to $nsidec/ntopc/ntipc$ where vertex j for the described concavity is stored.

Next, we will deal with the case of consecutive concavities in which one vertex will be common to both. This situation is possible for locally non-convex polygons as well as for the general case. As can be observed in the data structures, after writing the concavity vertices, we leave one component with a null value to delimit the end of the concavity. This component will serve to solve this case. Relying on the fact that in order for a vertex to be common to two concavities, it must be the last vertex of one of them and the first vertex of the other, this component will store the index that will allow us to link both concavities.

When the last edge of the first concavity or the first edge of the second concavity is cut in order to obtain a new representation, the conflict disappears. The next algorithm summarizes the steps to be followed in such case, beginning with the fact that an edge defined by vertices v_i and v_j that makes up part of a concavity has been cut.

$nsidep1 \leftarrow nconvex$ value for v_i
$nsidep2 \leftarrow nconvex$ value for v_j
$nvert \leftarrow$ number of vertices of the concavity
$norder1 \leftarrow nsidec(1, nsidep1)$, for v_i
$norder2 \leftarrow nsidec(1, nsidep2)$, for v_j

If (*norder1* > *norder2*) then
   v_i belongs to two concavities
   v_i is in *nsidec*(2, *nsidep1* + 1)
   nsidec(2, *nsidep1* + 1) ← 0
endif
If (*norder2* =*nvert*) then
   v_j is the last vertex of the concavity
   if (*nsidec*(2, *nsidep2* + 1)≠0) then
      v_j belongs to two concavities
      nconvex(v_j) ← *nsidec*(2, *nsidep2* + 1)
      nsidec(2, *nsidep2* + 1) ← 0
   endif
endif

We can see some of these actions for the example in Figure A.4 by comparing Figures A.5 and A.6, where the boxes that have been modified are shaded.

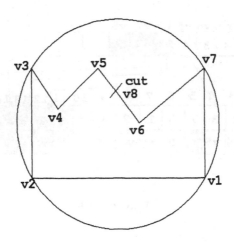

Figure   A.4

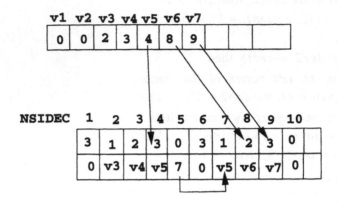

Figure A.5

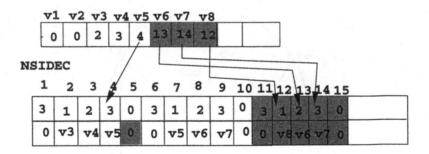

Figure A.6

References

Abramowski, S., 1988, "Collision Avoidance for Nonrigid Objects," in Computational Geometry and its Applications, ed. by H. Noltemeier, Springer-Verlag, Berlin, pp. 168-179.

Agin, G.J., Binford, T.O., 1976, "Computer Description of Curved Objects", IEEE Trans. on Computers Vol. C-25 No. 4, pp. 439-449.

Ahuja, N., Chien, R.T., Yen, R., Birdwell, N., 1980, "Interference Detection and Collision Avoidance Among Three-Dimensional Objects", Proc. First Annual Conference on Artificial Intelligence, Palo Alto, pp. 44-48.

Akman, V., 1985, "Unobstructed Shortest Paths in Polyhedral Environments, Lecture Notes in Computer Science Vol. 251, Springer-Verlag, Berlin. Also Ph. D. Thesis, Electrical, Computer and Systems Eng. Dept, Rensselaer Polytechnic Institute, Troy, New York, June 1985.

Albus, J., 1984, "Robots", in Brady M. et al (eds), Robotics and Artificial Intelligence, Springer-Verlag, Berlin, pp. 65-95.

Avis, D., Bhattacharya, B.K., Imai, H., 1983, "Computing the Volume of the Union of Spheres", The Visual Computer, Vol. 3, pp. 323-328.

Aumüller, F., Robinson, J.H., 1984, Freeston, S., "A Practical Exact Motion Planning Algorithm for Polyhedral Objects amidst Polygonal Obstacles", Proc. IEEE Conf. on Robotics and Automation, pp. 1458-1461.

References

Abramowski, S., 1988, "Collision Avoidance for Nonrigid Objects", in *Computational Geometry and its Applications*, ed. by H. Noltemeier, Springer-Verlag, Berlin, pp. 168-179.

Agin, G.J., Binford, T.O., 1976, "Computer Description of Curved Objects", *IEEE Trans. on Computers*, Vol. **C-25**, No. 4, pp. 439-449.

Ahuja, N., Chien, R.T., Yen, R., Birdwell, N., 1980, "Interference Detection and Collision Avoidance Among Three-Dimensional Objects", Proc. First Annual Conference on Artificial Intelligence, Palo Alto, pp. 44-48.

Akman, V., 1987, *Unobstructed Shortest Paths in Polyhedral Environments*, Lecture Notes in Computer Science Vol. 251, Springer-Verlag, Berlin. Also Ph. D. Thesis, Electrical, Computer and Systems Eng. Dept., Rensselaer Polytechnic Institute, Troy, New York, June 1985.

Albus, J., 1984, "Robotics", in Brady, M. et al. (eds.), *Robotics and Artificial Intelligence*, Springer-Verlag, Berlin, pp. 65-96.

Avis, D., Bhattacharya, B.K., Imai, H., 1988, "Computing the Volume of the Union of Spheres", *The Visual Computer*, Vol. **3**, pp. 323-328.

Avnaim, F., Boissonat, J.D., 1988, Faverjon, B., "A Practical Exact Motion Planning Algorithm for Polygonal Objects amidst Polygonal Obstacles", Proc. IEEE Intl. Conf. on Robotics and Automation, pp. 1656-1661.

Badler, N.I., O'Rourke, J., Toltzis, H., 1979, "A Spherical Representation of a Human Body for Visualizing Movement", *Proceeding of the IEEE*, Vol. **67**, No. 10, pp. 1397-1403.

Baer, A., Eastman, C., Henrion, M., 1979, "Geometric Modelling: A Survey", *Computer-Aided Design*, Vol. **11**, No. 5, pp. 253-272.

Bajaj, C., Kim, M.S., 1988, "Generation of Configuration Space Obstacles: The Case of a Moving Sphere", *IEEE J. on Robotics and Automation*, Vol. **4**, No. 1, pp. 94-99.

Bambah, R.P., 1954, "On Lattice Coverings by Spheres", *Proc. Nat. Inst. Sci. India*, Vol. **20**, pp. 25-52.

Barnes, E.S., 1956, "The Covering of Space by Spheres", *Canad. J. Math.*, Vol. **8**, pp. 293-304.

Bentley, J.L., Ottmann, T.A., 1979, "Algorithms for Reporting and Counting Geometric Intersections", *IEEE Trans. on Computers*, Vol. **C-28**, No. 9, pp. 642-647.

Bhattacharya, B.K., Toussaint, G.T., 1988, "Fast Algorithms for Computing the Diameter of a Planar Finite Set", *The Visual Computer*, Vol. **3**, pp. 379-388.

Bobrow, J.E., 1989, "A Direct Minimization Approach for Obtaining the Distance Between Convex Polyhedra", *The Intl. J. of Robotics Research*, Vol. **8**, No. 3, pp. 65-76.

Bonner, S., Kelley, R.B., 1990, "A Novel Representation for Planning 3-D Collision-Free Paths", *IEEE Trans. on Systems, Man, and Cybernetics*, Vol. **20**, No. 6 , pp.1337-1351.

Boyse, J.W., 1979, "Interference Detection Among Solids and Surfaces", *Comm. of the ACM*, Vol. **22**, No. 1, pp. 3-9.

Brady, M. (ed.), 1989, *Robotics Science*, The MIT Press, Cambridge, Massachusetts.

Brady, M. et al. (eds.), 1982, *Robot Motion: Planning and Control,* The MIT Press, Cambridge, Massachusetts.

Brady, M. et al., (eds.), 1984, *Robotics and Artificial Intelligence,* Springer-Verlag, Berlin.

Brady, M., 1984, "Representing Shape", in Brady, M. et al., (eds.) *Robotics and Artificial Intelligence,* Springer-Verlag, Berlin, pp. 279-300.

Brady, M., 1985, "Artificial Intelligence and Robotics", *Artificial Intelligence,* Vol. **26**, pp. 79-121.

Brooks, R.A., 1982, "Symbolic Error Analysis and Robot Planning", *Intl. J. Robotics Research,* Vol. **1**, No. 4, pp. 21-68.

Brooks, R.A., 1983a, "Solving the Find-Path Problem by Good Representation of Free Space", *IEEE Trans. on Systems, Man and Cybernetics,* Vol. **SMC-13**, No. 3, pp. 190-197.

Brooks, R.A., 1983b, "Planning Collision-Free Motions for Pick-and-Place Operations", *Intl. J. Robotics Research,* Vol. **2**, No. 4, pp. 19-44.

Brooks, R.A., Lozano-Pérez, T., 1985, "A Subdivision Algorithm in Configuration Space for Findpath with Rotation", *IEEE Trans. on Systems, Man and Cybernetics,* Vol. **SMC-15**, No. 2, pp. 224-233.

Brost, R.C., 1986, "Automatic Grasp Planning in the Presence of Uncertainty", Proc. IEEE Intl. Conf. on Robotics and Automation, pp. 1575-1581.

Buckley, C.E., 1989a, "A Foundation for the 'Flexible-Trajectory' Approach to Numeric Path Planning", *The Intl. J. of Robotics Research,* Vol. **8**, No. 3, pp. 44-64.

Buckley, S.J., 1989b, "Fast Motion Planning for Multiple Moving Robots", Proc. IEEE Intl. Conf. on Robotics and Automation, pp. 322-326.

Buchanan, B.G., Shortliffe, E.H., 1984, *Rule-Based Expert Systems*, Addison-Wesley, Reading, Massachusetts.

Cameron, S., 1985, "A Study of the Clash Detection Problem in Robotics", Proc. IEEE Intl. Conf. on Robotics and Automation, pp. 488-493.

Cameron, S., 1989, "Efficient Intersection Tests for Objects Defined Constructively", *Intl. J. Robotics Research*, Vol. **8**, pp. 3-25.

Cameron, S., 1990, "Collision Detection by Four-Dimensional Intersection Testing", *IEEE Trans. on Robotics and Automation*, Vol. **6**, No. 3, pp. 291-302.

Canny, J., 1986, "Collision Detection for Moving Polyhedra", *IEEE Trans. on Patt. Anal. & Mach. Intell.*, Vol. **PAMI-8**, No.2, pp. 200-209.

Canny, J., 1988, *The Complexity of Robot Motion Planning*, The MIT Press, Cambridge, Massachusetts.

Canny, J., Donald, B., 1988, "Simplified Voronoi Diagrams", *Discrete and Computational Geometry*, Vol. **3**, No. 3, pp. 219-236. Also in Cox, I.J., Wilfong, G.T. (eds.), *Autonomous Robot Vehicles*, Springer-Verlag, New York, 1990, pp. 272-289.

Canny, J., Lin, M.C., 1990, "An Opportunistic Global Path Planner", Proc. IEEE Intl. Conf. on Robotics and Automation, pp. 1554-1560.

Connolly, M.L., 1987, "An Application of Algebraic Topology to Solid Modeling in Molecular Biology", *The Visual Computer*, Vol. **3**, pp. 72-81.

Cox, I.J., Wilfong, G.T. (eds.), 1990, *Autonomous Robot Vehicles*, Springer-Verlag, Berlin.

Coxeter, H.S.M., Few, L., Rogers, C.A., 1959, "Covering Space with Equal Spheres", *Mathematika*, Vol. **6**, pp. 147-157.

Charniak, E., McDermott, D., 1985, *Introduction to Artificial Intelligence*, Addison-Wesley, Reading Massachusetts.

Chazelle, B., 1987, "Approximation and Decomposition of Shapes", in *Algorithmic and Geometric Aspects of Robotics*, ed. by J.T. Schwartz and C.K. Yap, Lawrence Erlbaum, Chap. 4, pp. 145-185.

Chazelle, B., Edelsbrunner, H., 1988, "An Optimal Algorithm for Intersecting Line Segments in the Plane", University of Illinois at Urbana-Champaign, Dept. of Computer Science, Report UIUCDCS-R-88-1419.

Chen, S., 1990, "A Spherical Model for Navigation and Spatial Reasoning", Proc. IEEE Intl. Conf. on Robotics and Automation, pp. 776-781.

Chen, Y.C., Vidyasagar, M., 1987, "Some Qualitative Results on the Collision-Free Joint Space of a Planar n-DOF Linkage", Proc. IEEE Intl. Conf. on Robotics and Automation, pp. 1623-1629.

Chien, R.T., Zhang, L., Zhang, B., 1984, "Planning Collision-Free Paths for Robotic Arm Among Obstacles", *IEEE Trans. on Pattern Analysis and Machine Intelligence*, Vol. **PAMI-6**, No. 1, pp. 91-96

Davis, E., 1988, "A Logical Framework for Commonsense Predictions of Solid Object Behaviour", *Artificial Intelligence in Engineering*, Vol. **3**, No. 3, pp.125-140.

Davis, E., 1990, *Representations of Commonsense Knowledge*, Morgan Kaufmann Publishers, San Mateo, CA.

de Pennington, A., Bloor, M.S., Balila, M., 1983, "Geometric Modelling: A Contribution Towards Intelligent Robots", Proc. 13th Intl. Symposium on Industrial Robots, Chicago, pp. 7.35-7.54.

del Pobil, A.P., Escrig, M.T., Jaén, J.A., 1993, "An Attempt Towards a General Representation Paradigm for Spatial Reasoning", *Proc. IEEE International Conference on Systems, Man and Cybernetics*, Le Touquet, France, vol. 1, pp. 215-220.

del Pobil, A.P., Escrig, T., Jaén, J.A., 1994, "Building Qualitative Spatial Relations from Bottom-Up", *Proc. 14th IMACS World Congress*, Atlanta, Georgia, pp. 1416-1419.

del Pobil, A.P., Muñoz, C., García, J.L., 1988, "Implementation du systeme expert Iroise dans un contexte multitâche en mode serveur sur reseau Ethernet", AMAIA-Bayonne Systèmes Informatiques.

del Pobil, A.P., Serna, M.A., 1992, "Solving the Find-Path Problem by a Simple Object Model", *Proc. 10th European Conference on Artificial Intelligence ECAI-92*, Vienna, Austria, pp. 656-660.

del Pobil, A.P., Serna, M.A., 1994a, "A New Object Representation for Robotics and Artificial Intelligence Applications", *International Journal of Robotics & Automation*, Vol. 9, No. 1, pp. 11-21.

del Pobil, A.P., Serna, M.A., 1994b, "A Simple Algorithm for Intelligent Manipulator Collision-Free Motion", *Journal of Applied Intelligence*, Vol. 4, pp. 83-102.

del Pobil, A.P., Serna, M.A., Llovet, J., 1992, "A New Representation for Collision Avoidance and Detection", *Proc. IEEE International Conference on Robotics and Automation*, Nice, France, pp. 246-251.

Dobkin, D.P., Kirkpatric, D.G., 1985, "A Linear Algorithm for Determining the Separation of Convex Polyhedra", *Journal of Algorithms*, Vol. 6, pp. 381-392.

Donald, B.R., 1987, "A Search Algorithm for Motion Planning with Six Degrees of Freedom", *Artificial Intelligence*, Vol. 31, No. 3, pp. 295-353.

Donald, B.R., 1988, "A Geometric Approach to Error Detection and Recovery for Robot Motion Planning with Uncertainty", *Artificial Intelligence*, Vol. 37, pp. 223-271.

Dupont, P.E., 1988, "Collision-Free Path Planning for Kinematically Redundant Robots", Ph. D. Thesis, Mechanical Engineering Dept., Rensselaer Polytechnic Institute.

Edelsbrunner, H., 1982, "Intersection Problems in Computational Geometry", Ph. D. Thesis, Technical University of Graz (Austria), Institut für Informationverarbeitung, Report F-93.

Edelsbrunner, H., Seidel, R., 1985, "Voronoi Diagrams and Arrangements", Cornell University, Dept. of Computer Science, Technical Report TR-85-669. Also in Proc. ACM Symp. on Computational Geometryy, 1985, pp. 251-262.

Erdmann, M.A., 1984, "On Motion Planning with Uncertainty", Massachusetts Institute of Technology, Artificial Intelligence Lab., Technical Report 810.

Erdmann, M.A., Lozano-Pérez, T., 1987, "On Multiple Moving Objects", *Algorithmica*, Vol. 2, No. 4, pp. 477-521.

Erdmann, M.A., Mason, M.T., 1986, "An Exploration of Sensorless Manipulation", Proc. IEEE Intl. Conf. on Robotics and Automation, pp. 1569-1574.

Esterling, D.M., Van Rosendale, J., 1983, "An Intersection Algorithm for Moving Parts", Proc. NASA Symposium on Computer-Aided Geometry Modeling, Hampton, Virginia, pp. 119-123.

Faugeras, O.D., Hebert, M., Pauchon, E., Ponce, J., 1984, "Object Representation, Identification and Positioning from Range Data", in Brady, M. et al., (eds.) *Robotics and Artificial Intelligence*, Springer-Verlag, Berlin, pp. 255-277.

Faugeras, O.D., Ponce, J., 1983, "Prism-Trees: a Hierarchical Representation for 3-D Objects", Proc. 8th Intl. Joint Conference on Artificial Intelligence, pp. 982-988.

Faverjon, B., 1989, "Hierarchical Object Models for Efficient Anti-Collision Algorithms", Proc. IEEE Intl. Conf. on Robotics and Automation, pp. 333-340.

Faverjon, B., Tournassoud, P., 1986, "A Hierarchical CAD System for Multi-Robot Coordination", Proc. NATO Advanced Research Workshop on Languages for Sensor Based Control in Robotics, Pisa, Italy, pp. 317-327.

Faverjon, B., Tournassoud, P., 1988, "A Practical Approach to Motion-Planning for Manipulators with Many Degrees of Freedom", INRIA, Rapport de Recherche No. 951.

Few, L., 1956, "Covering Space by Spheres", *Mathematika*, Vol. **3**, pp. 136-39.

Francis, R.L., White, J.A., 1974, *Facility Layout and Location*, Prentice-Hall, Englewood Cliffs, New Jersey.

Fujimura, K., Samet, H., 1989, "A Hierarchical Strategy for Path Planning Among Moving Obstacles", *IEEE Trans. on Robotics and Automation*, Vol. **5**, No. 1, pp. 61-69.

García de Jalón, J., Unda, J., Avello, A., 1986, "Natural Coordinates for the Computer Analysis of Three-Dimensional Multibody Systems", *Computer Methods in Applied Mechanics and Engineering*, Vol. **56**, No. 3, pp. 309-327.

García-Alonso, A., Serrano, N., Flaquer, J., 1994, "Solving the Collision Detection Problem", *IEEE Computer Graphics and Applications*, Vol. **14**, No. 3, pp. 36-43.

Gilbert, E.G., Foo, C.P., 1990, "Computing the Distance Between General Convex Objects in Three-Dimensional Space", *IEEE Trans. on Robotics and Automation*, Vol. **RA-6**, pp. 53-61.

Gouzènes, L., 1984, "Strategies for Solving Collision-Free Trajectory Problems for Mobile and Manipulator Robots", *Intl. J. Robotics Research*, Vol. **3**, No. 4, pp. 51-65.

Hayes-Roth, F., Waterman, D.A., Lenat, D.B. (eds.), **1983,** *Building Expert Systems*, Addison-Wesley, Reading, Massachuetts.

Hayward, V., 1986, "Fast Collision Detection Scheme by Recursive Decomposition of A Manipulator Workspace", Proc. IEEE Intl. Conf. on Robotics and Automation, pp. 1044-1049.

Hollerbach, J.M., 1982, "Computer, Brains, and the Control of Movement", Massachusetts Institute of Technology, Artificial Intelligence Lab., Memo 686.

Hopcroft, J.E., Joseph, D., Whitesides, S., 1984, "Movement Problems for Two-Dimensional Linkages", *SIAM J. on Computing*, Vol. **13**, pp. 610-629.

Hopcroft, J.E., Joseph, D., Whitesides, S., 1985, "On the Movement of Robot Arms in 2-Dimensional Bounded Regions", *SIAM J. on Computing*, Vol. **14**, No. 2.

Hopcroft, J.E., Krafft, D.B., 1987, "The Challenge of Robotics for Computer Science", in *Advances in Robotics (Vol. I): Algorithmic and Geometric Aspects of Robotics*, ed. by J.T. Schwartz and C.K. Yap, Lawrence Erlbaum, Hillsdale, New Jersey, Chap. 1, pp. 7-42.

Hopcroft, J.E., Schwartz, J.E., Sharir, M., 1983, "Efficient Detection of Intersections Among Spheres", *Intl. J. Robotics Research*, Vol. **2**, No. 4, pp. 77-80.

Hopcroft, J.E., Schwartz, J.E., Sharir, M., 1984, "On the Complexity of Motion Planning for Multiple Independent Objects; PSPACE-Hardness of the 'Warehouseman´s Problem'", *Intl. J. Robotics Research*, Vol. **3**, No. 4, pp. 76-88.

Hopcroft, J.E., Wilfong, G., 1986, "Motion of Objects in Contact", *Intl. J. Robotics Research*, Vol. **4**, No. 4, pp. 32-46.

Huet, G., 1986, "Manuel de realisation Iroise", Centre National d'Etudes des Telecommunications, Division Structures et Logiciels par la Commutation, Note Technique NT/LAA/SLC/230.

Hwang, Y.K., Ahuja, N., 1992, "Gross Motion Planning. A Survey", *ACM Computing Surveys*, Vol. 24, No. 3, pp. 219-291.

Ichikawa, Y., Ozaki, N., 1985, "Autonomous Mobile Robot", *J. of Robotic Systems*, Vol. 2, No. 1, pp. 135-144.

Jacobs, P., Canny, J., 1990, "Robust Motion Planning for Mobile Robots", Proc. IEEE Intl. Conf. on Robotics and Automation, pp. 2-7.

Juan, J., Paul, R.P., 1986, "Automatic Programming of Fine-Motion for Assembly", Proc. IEEE Intl. Conf. on Robotics and Automation, pp. 1582-1587.

Kambhampati, S., Davis, L.S., 1986, "Multiresolution Path Planning for Mobile Robots", *IEEE J. of Robotics and Automation*, Vol. RA-2, No. 3, pp. 135-145.

Kant, K., Zucker, S.W., 1988, "Planning Collision-Free Trajectories in Time-Varying Environments: a Two Level Hierarchy", *The Visual Computer*, Vol. 2, pp. 304-313.

Kantabutra, V., Kosaraju, S.R., 1986, "New Algorithms for Multilink Robot Arms", *J. of Computer and System Science*, Vol. 32, pp. 136-153.

Kariv, O., Hakimi, S.L., 1979, "An Algorithmic Approach to Network Location Problems, Part I: The p-centers", *SIAM J. Appl. Math.* Vol. 37, pp. 513-538.

Kawabe, S., Okano, A., Shimada, K., 1988, "Collision Detection Among Moving Objects in Simulation", in Bolles, R., Roth, B. (eds.), *Robotics Research*, The MIT Press, Cambridge, Massachusetts, pp. 489-496.

Kedem, K., Sharir, M.,, 1986, "An Efficient Motion Planning Algorithm for a Convex Polygonal Object in 2-Dimensional Polygonal Space", New York University, Courant Institute of Mathematical Sciences, Technical Report No. 253.

Kershner, R., 1939, "The number of circles covering a set", *Amer. J. Math.*, Vol. **61**, pp. 665-671.

Khatib, O., 1986, "Real Time Obstacle Avoidance for Manipulators and Mobile Robots", *Intl. J. Robotics Research*, Vol. **5**, No. 1, pp. 90-98.

Khatib, O., Craig, J.J., 1989, Lozano-Pérez, T. (eds.), *The Robotics Review 1*, The MIT Press, Cambridge, Massachusetts.

Khosla, P., Volpe, R., 1988, "Superquadric Artificial Potentials for Obstacle Avoidance and Approach", Proc. IEEE Intl. Conf. on Robotics and Automation, pp. 1778-1784.

Klein, R., 1988, "Abstract Voronoi Diagrams and their Applications", in Noltemeier, H. (ed.), *Computational Geometry and its Applications*, Springer-Verlag, Berlin, pp. 148-157.

Korein, J.U., 1985, *A Geometric Investigation of Reach*, The MIT Press, Cambridge, Massachusetts. Also Ph. D. Thesis, University of Pennsylvania, Dept. of Computer and Information Science, Feb. 1984.

Korein, J.U., Ish-Shalom, J., 1987, "Robotics", *IBM Systems Journal*, Vol. **26**, No. 1, pp. 55-95.

Latombe, J.C, Lazanas, A., Shekhar, S., 1991, "Robot Motion Planning with Uncertainty in Control and Sensing", *Artificial Intelligence*, Vol. **53**, pp. 1-47.

Latombe, J.C., 1991, *Robot Motion Planning*, Kluwer Academic Publishers, Boston.

Lee, B.H., Lee, C.S.G., 1987, "Collision-Free Motion Planning of Two Robots", *IEEE Trans. on Systems, Man and Cybernetics*, Vol. **SMC-17**, No. 1, pp. 21-32.

Lee, D.T., Drysdale, R.L., 1981, "Generalization of Voronoi Diagrams in the Plane", *SIAM J. on Computing*, Vol. **10**, No. 1, pp. 73-87.

Lee, D.T., Preparata, F.P., 1984, "Computational Geometry: A Survey", *IEEE Trans. on Computers*, Vol. **C-33**, No. 12, pp. 1072-1101.

Lee, Y.C., Fu, K.S., 1987, "Machine Understanding of CSG: Extraction and Unification of Manufacturing Features", *IEEE Computer Graphics & Applications*, pp. 20-32.

Leven, D., Sharir, M., 1987a, "An Efficient and Simple Motion Planning Algorithm for a Ladder Amidst Polygonal Barriers", *Journal of Algorithms*, Vol. **8**, pp. 192-215.

Leven, D., Sharir, M., 1987b, "Intersection and Proximity Problems and Voronoi Diagrams", in Schwartz, J.T., Yap, C.K. (eds.), *Advances in Robotics (Vol. I): Algorithmic and Geometric Aspects of Robotics*, Lawrence Erlbaum, Hillsdale, New Jersey, pp. 187-228.

Lozano-Pérez, T., 1983, "Spatial Planning: A Configuration Space Approach", *IEEE Trans. on Computers*, Vol. **C-32**, No. 2, pp. 108-120.

Lozano-Pérez, T., 1987, "A Simple Motion-Planning Algorithm for General Robot Manipulators", *IEEE J. on Robotics and Automation*, Vol. **RA-3**, No. 3, pp. 224-238.

Lozano-Pérez, T., Mason, M.T., Taylor, R.H., 1984, "Automatic Synthesis of Fine-Motion Strategies for Robots", *Intl. J. Robotics Research*, Vol. **3**, No. 1, pp. 3-24.

Lozano-Pérez, T., Wesley, M.A., 1979, "An algorithm for Planning Collision-Free Paths Among Polyhedral Obstacles", *Comm. of the ACM*, Vol. **22**, No. 10, pp. 560-570.

Maciejewski, A.A., Klein, C.A., 1985, "Obstacle Avoidance for Kinematically Redundant Manipulators in Dynamically Varying Environments", *The Intl. J. of Robotics Research*, Vol. **4**, No. 3, pp. 109-117.

Marr, D., Vaina, L., 1980, "Representation and Recognition of the Movements of Shapes", Massachusetts Institute of Technology, Artificial Intelligence Lab., Memo 597.

Martín, P., del Pobil, A.P., 1994a, "A Connectionist System for Learning Robot Manipulator Obstacle-Avoidance Capabilities in Path-Planning", *Proc. IMACS International Symposium on Signal Processing, Robotics and Neural Networks*, Lille, France, pp. 141-144.

Martín, P., del Pobil, A.P., 1994b, "Application of Artificial Neural Networks to the Robot Path Planning Problem", in *Applications of Artificial Intelligence in Engineering IX* edited by G. Rzevski, R.A. Adey and D.W. Russell, Computational Mechanics Publications, Boston, pp. 73-80.

Mason, M.T., 1982, "Compliance", in Brady, M. et al. (eds.), *Robot Motion: Planning and Control*, The MIT Press, Cambridge, Massachusetts, pp. 305-322.

Mason, M.T., 1984, "Automatic Planning of Fine Motions: Correctness and Completeness", Proc. IEEE Intl. Conf. on Robotics and Automation,.

McCreight, E.M., 1985, "Priority Search Trees", *SIAM J. on Computing*, Vol. **14**, No. 2, pp. 257-275.

McKerrow, P.J., 1991, *Introduction to Robotics*, Addison-Wesley, Sidney, 1991.

Megiddo, N., 1983, "Linear Time Algorithm for Linear Programming in $\Re^3$ and Related Problems", *SIAM J. Computing*, Vol. **12**, No. 4, pp. 759-776.

Megiddo, N., Supowit, K.J., 1984, "On the Complexity of Some Common Geometric Location Problems", *SIAM J. Computing*, Vol. **13**, No. 1, pp. 182-196.

Meyer, J., 1981, "An Emulation System for Programmable Sensory Robots", *IBM Journal of Research & Development*, Vol. **25**, No. 6, pp. 955-962.

Milenkovic, V., Huang, B., 1983, "Kinematics of Major Robot Linkage", Proc. 13th Intl. Symp. on Industrial Robots, pp. 16.31-16.47, Chicago.

Moore, M., Wilhelms, J., 1988, "Collision Detection and Response for Computer Animation", *Computer Graphics*, Vol. **22**, No. 4, pp. 289-298.

Moravec, H.P., 1981, "Rover Visual Obstacle Avoidance", Proc. 7th Intl. Joint Conference on Artificial Intelligence, pp. 785-790.

Nguyen, V.D., 1984, "The Find-Path Problem in the Plane", Massachusetts Institute of Technology, Artificial Intelligence Lab., Memo 760.

O'Dúnlaing, C., 1987, "Motion Planning with Inertial Constraints", *Algorithmica*, Vol. **2**, pp. 431-475.

O'Dúnlaing, C., Sharir, M., Yap, C.K., 1983, "Retraction: A New Approach to Motion Planning", Proc. 15th ACM Sympos. Theory of Computing, pp. 207-220.

O'Dúnlaing, C., Sharir, M., Yap, C.K., 1986, "Generalized Voronoi Diagrams for Moving a Ladder. I: Topological Analysis", *Comm. on Pure and Applied Mathematics*, Vol. **34**, pp. 423-483.

O'Dúnlaing, C., Sharir, M., Yap, C.K., 1987, "Generalized Voronoi Diagrams for a Ladder. II: Efficient Construction of the Diagram", *Algorithmica*, Vol. **2**, pp. 27-59.

O'Dúnlaing, C., Yap, C.K., 1985, "A "Retraction" Method for Planning the Motion of a Disc", *Journal of Algorithms*, Vol. **6**, pp. 104-111.

Ozaki, H., 1986, "On the Planning of Collision-Free Movements of Manipulators", *Advanced Robotics*, Vol. **1**, No. 3, pp. 261-272.

Parsons, D., Canny, J., 1990, "A Motion Planner for Multiple Mobile Robots", Proc. IEEE Intl. Conf. on Robotics and Automation, pp. 8-13.

Paul, R.P., 1981, *Robot Manipulators: Mathematics, Programming and Control*, The MIT Press, Cambridge, Massachusetts.

Pearl, J., 1984, *Heuristics: Intelligent Search Strategies for Computer Problem Solving*, Addison-Wesley, Reading, Massachusetts.

Pieper, D.L., 1968, "The Kinematics of Manipulators under Computer Control", Ph.D. Thesis, Stanford University.

Preparata, F.P., Shamos, M.I., 1985., *Computational Geometry. An Introduction*, Springer-Verlag, New York.

Rao, K., Nevaitia, R., 1990, "Computing Volume Descriptions from Sparse 3-D Data", in *Advances in Spatial Reasoning*, edited by S. Chen, Ablex, Norwood, New Jersey.

Reddy, D.R., Rubin, S., 1978, "Representation of Three-Dimensional Objects", Carnegie-Mellon University, Dept. of Computer Science, Report CMU-CS-78-113.

Reif, J., 1979, "Complexity of the Mover´s Problem and Generalizations", Proc. Symp. on the Foundations of Computer Science, pp. 421-427.

Requicha, A., 1980, "Representations for Rigid Solids: Theory, Methods and Systems", *Computing Surveys*, Vol. 12, No. 4, pp. 437-465.

Rogers, C.A., 1964, *Packing and Covering*, Cambridge University Press, Cambridge, England.

Samet, H., 1984, "The Quadtree and Related Hierarchical Data Structures", *ACM Computing Surveys*, Vol. 16, No. 2, pp. 187-260.

Schwartz, J., Hopcroft, J., Sharir, M. (eds.), 1987, *Planning, Geometry and Complexity of Robot Motion*, Ablex Publishing Corp., Norwood, New Jersey.

Schwartz, J., Sharir, M., 1983a, "On the Piano Movers' Problem: II. General Techniques for Computing Topological Properties of Real

Algebraic Manifolds", *Advances in Applied Mathematics*, Vol. **4**, pp. 298-351.

Schwartz, J., Sharir, M., **1983b**, "On the Piano Movers' Problem: III. Coordinating the Motion of Several Independent Bodies: The Special Case of Circular Bodies Moving Amidst Polygonal Barriers", *Intl. J. Robotics Research*, Vol. **2**, No. 3, pp. 46-75.

Schwartz, J., Sharir, M., **1984**, "On the Piano Movers' Problem: V. The Case of a Rod Moving in Three-Dimensional Space Amidst Polyhedral Obstacles. *Comm. on Pure and Applied Mathematics*, Vol. **37**, pp. 815-848.

Schwartz, J., Sharir, M., **1988**, "A Survey of Motion Planning and Related Geometric Algorithms", *Artificial Intelligence*, Vol. **37**, pp. 157-169.

Shafer, S.A., **1983**, "Shadow Geometry and Occluding Contours of Generalized Cylinders", Tech. Report No. CS-83-131, Carnegie-Mellon University, Pittsburgh, Pennsylvania.

Shamos, M.I., Hoey, D., **1975**, , "Closest Point Problems", Proc. 16th IEEE Annu. Symp. Found. Comput. Sci. pp. 151-162.

Shamos, M.I., Hoey, D., **1976**, "Geometric Intersection Problems", Proc. 17th IEEE Annual Symposium on Foundations of Computer Science, Houston, pp. 208-215.

Shannon, R.E., Ignizio, J.P., **1970**, "A Heuristic Programming Algorithm for Warehouse Location", *AIIE Transactions*, Vol. **2**, No. 4, pp. 334-339.

Sharir, M., **1985**, "Intersection and Closest-Pair Problems for a Set of Planar Discs", *SIAM J. on Computing*, Vol. **14**, No. 2, pp. 448-468.

Sharir, M., **1987**, "Efficient Algorithms for Planning Purely Translational Collision-free Motion in Two and Three Dimensions", Proc. IEEE Intl. Conf. on Robotics and Automation, pp. 1326-1331.

Shih, C.L., Lee, T.T., Gruver, W.A., 1990, "A Unified Approach for Robot Motion Planning with Moving Polyhedral Obstacles", *IEEE Trans. on Systems, Man and Cybernetics,* Vol. **SMC-20**, No. 4, pp. 903-915.

Singh, S.K., 1988, "Motion Planning with obstacles and dynamic constraints", Ph.D. Thesis, Cornell University.

Spirakis, P., Yap, C., 1984, "Strong NP-Hardness of Moving Many Discs", *Information Processing Letters*, Vol. **19**, pp. 55-59.

Takahashi, O., Schilling, J., 1989, "Motion Planning in a Plane Using Generalized Voronoi Diagrams", *IEEE Trans. on Robotics and Automation,* Vol. **5**, No. 2, pp. 143-150.

Thakur, B.K., 1986, "Automatic Path-Planning of Industrial Robots", Ph.D. Thesis, Rensselaer Polytechnic Institute.

Thorpe, C.E., 1984, "Path Relaxation: Path Planning for a Mobile Robot", Carnegie-Mellon University, Robotics Institute, Technical Report CMU-RI-TR-84-5.

Tornero, J., Hamlin, J., Kelley, R.B., 1991, "Spherical-Object Representation and Fast Distance Computation for Robotic Applications", Proc. IEEE Intl. Conf. on Robotics and Automation, pp. 1602-1608.

Tournassoud, P., Jehl, O., 1988, "Motion Planning for a Mobile Robot with a Kinematic Constraint", Proc. IEEE Intl. Conf. on Robotics and Automation, pp. 1785-1790.

Uchiki, T., Ohashi, T., Tokoro, M., 1983, "Collision Detection in Motion Simulation", *Comput. & Graphics*, Vol. **7**, No. 3-4, pp. 285-293.

Voronoï, G., 1908, "Nouvelles applications des paramètres continus à la theorie des formes quadratiques", Deuxième Mémoire, Recherches sur les parallélloèdres primitifs, *J. reine angew. Math.*, Vol. **134**, pp. 198-287.

Wang, W.P., Wang, K.K., 1986, "Geometric Modeling for Swept Volume of Moving Solids", *IEEE Computer Graphics & Applications*, pp. 8-17.

Widdoes, C., 1974, "A Heuristic Collision Avoider for the Stanford Robot Arm", Stanford C.S. Memo 227.

Wong, E.K., Fu, K.S., 1985, "A Hierarchical-Orthogonal-Space Approach to Collision-Free Path Planning", Proc. IEEE Intl. Conf. on Robotics and Automation, pp. 506-511.

Yap, C.K., 1984, "Coordinating the Motion of Several Discs", New York University, Courant Institute of Mathematical Sciences, Technical Report No. 105.

Yap, C.K., 1987, "Algorithmic Motion Planning", in Schwartz, J.T., Yap, C.K. (eds.), *Advances in Robotics (Vol. I): Algorithmic and Geometric Aspects of Robotics*, Lawrence Erlbaum, Hillsdale, New Jersey, pp. 95-144.

Zhu, D., Latombe, J.C., 1991, "New Heuristic Algorithms for Efficient Hierarchical Path Planning", *IEEE Trans. on Robotics and Automation*, Vol. 7, No. 1, pp. 9-20.

Lecture Notes in Computer Science

For information about Vols. 1–935

please contact your bookseller or Springer-Verlag

Vol. 971: E.T. Schubert, P.J. Windley, J. Alves-Foss (Eds.), Higher Order Logic Theorem Proving and Its Applications. Proceedings, 1995. VIII, 400 pages. 1995.

Vol. 972: J.-M. Hélary, M. Raynal (Eds.), Distributed Algorithms. Proceedings, 1995. XI, 333 pages. 1995.

Vol. 973: H.H. Adelsberger, J. Lažanský, V. Mařík (Eds.), Information Management in Computer Integrated Manufacturing. IX, 665 pages. 1995.

Vol. 974: C. Braccini, L. DeFloriani, G. Vernazza (Eds.), Image Analysis and Processing. Proceedings, 1995. XIX, 757 pages. 1995.

Vol. 975: W. Moore, W. Luk (Eds.), Field-Programmable Logic and Applications. Proceedings, 1995. XI, 448 pages. 1995.

Vol. 976: U. Montanari, F. Rossi (Eds.), Principles and Practice of Constraint Programming — CP '95. Proceedings, 1995. XIII, 651 pages. 1995.

Vol. 977: H. Beilner, F. Bause (Eds.), Quantitative Evaluation of Computing and Communication Systems. Proceedings, 1995. X, 415 pages. 1995.

Vol. 978: N. Revell, A. M. Tjoa (Eds.), Database and Expert Systems Applications. Proceedings, 1995. XV, 654 pages. 1995.

Vol. 979: P. Spirakis (Ed.), Algorithms — ESA '95. Proceedings, 1995. XII, 598 pages. 1995.

Vol. 980: A. Ferreira, J. Rolim (Eds.), Parallel Algorithms for Irregularly Structured Problems. Proceedings, 1995. IX, 409 pages. 1995.

Vol. 981: I. Wachsmuth, C.-R. Rollinger, W. Brauer (Eds.), KI-95: Advances in Artificial Intelligence. Proceedings, 1995. XII, 269 pages. (Subseries LNAI).

Vol. 982: S. Doaitse Swierstra, M. Hermenegildo (Eds.), Programming Languages: Implementations, Logics and Programs. Proceedings, 1995. XI, 467 pages. 1995.

Vol. 983: A. Mycroft (Ed.), Static Analysis. Proceedings, 1995. VIII, 423 pages. 1995.

Vol. 984: J.-M. Haton, M. Keane, M. Manago (Eds.), Advances in Case-Based Reasoning. Proceedings, 1994. VIII, 307 pages. 1995.

Vol. 985: T. Sellis (Ed.), Rules in Database Systems. Proceedings, 1995. VIII, 373 pages. 1995.

Vol. 986: Henry G. Baker (Ed.), Memory Management. Proceedings, 1995. XII, 417 pages. 1995.

Vol. 987: P.E. Camurati, H. Eveking (Eds.), Correct Hardware Design and Verification Methods. Proceedings, 1995. VIII, 342 pages. 1995.

Vol. 988: A.U. Frank, W. Kuhn (Eds.), Spatial Information Theory. Proceedings, 1995. XIII, 571 pages. 1995.

Vol. 989: W. Schäfer, P. Botella (Eds.), Software Engineering — ESEC '95. Proceedings, 1995. XII, 519 pages. 1995.

Vol. 990: C. Pinto-Ferreira, N.J. Mamede (Eds.), Progress in Artificial Intelligence. Proceedings, 1995. XIV, 487 pages. 1995. (Subseries LNAI).

Vol. 991: J. Wainer, A. Carvalho (Eds.), Advances in Artificial Intelligence. Proceedings, 1995. XII, 342 pages. 1995. (Subseries LNAI).

Vol. 992: M. Gori, G. Soda (Eds.), Topics in Artificial Intelligence. Proceedings, 1995. XII, 451 pages. 1995. (Subseries LNAI).

Vol. 993: T.C. Fogarty (Ed.), Evolutionary Computing. Proceedings, 1995. VIII, 264 pages. 1995.

Vol. 994: M. Hebert, J. Ponce, T. Boult, A. Gross (Eds.), Object Representation in Computer Vision. Proceedings, 1994. VIII, 359 pages. 1995.

Vol. 995: S.M. Müller, W.J. Paul, The Complexity of Simple Computer Architectures. XII, 270 pages. 1995.

Vol. 996: P. Dybjer, B. Nordström, J. Smith (Eds.), Types for Proofs and Programs. Proceedings, 1994. X, 202 pages. 1995.

Vol. 997: K.P. Jantke, T. Shinohara, T. Zeugmann (Eds.), Algorithmic Learning Theory. Proceedings, 1995. XV, 319 pages. 1995.

Vol. 998: A. Clarke, M. Campolargo, N. Karatzas (Eds.), Bringing Telecommunication Services to the People – IS&N '95. Proceedings, 1995. XII, 510 pages. 1995.

Vol. 999: P. Antsaklis, W. Kohn, A. Nerode, S. Sastry (Eds.), Hybrid Systems II. VIII, 569 pages. 1995.

Vol. 1000: J. van Leeuwen (Ed.), Computer Science Today. XIV, 643 pages. 1995.

Vol. 1004: J. Staples, P. Eades, N. Katoh, A. Moffat (Eds.), Algorithms and Computation. Proceedings, 1995. XV, 440 pages. 1995.

Vol. 1005: J. Estublier (Ed.), Software Configuration Management. Proceedings, 1995. IX, 311 pages. 1995.

Vol. 1006: S. Bhalla (Ed.), Information Systems and Data Management. Proceedings, 1995. IX, 321 pages. 1995.

Vol. 1007: A. Bosselaers, B. Preneel (Eds.), Integrity Primitives for Secure Information Systems. VII, 239 pages. 1995.

Vol. 1008: B. Preneel (Ed.), Fast Software Encryption. Proceedings, 1994. VIII, 367 pages. 1995.

Vol. 1009: M. Broy, S. Jähnichen (Eds.), KORSO: Methods, Languages, and Tools for the Construction of Correct Software. X, 449 pages. 1995. Vol.

Vol. 1010: M. Veloso, A. Aamodt (Eds.), Case-Based Reasoning Research and Development. Proceedings, 1995. X, 576 pages. 1995. (Subseries LNAI).

Vol. 1011: T. Furuhashi (Ed.), Advances in Fuzzy Logic, Neural Networks and Genetic Algorithms. Proceedings, 1994. (Subseries LNAI).

Vol. 1012: M. Bartošek, J. Staudek, J. Wiedermann (Eds.), SOFSEM '95: Theory and Practice of Informatics. Proceedings, 1995. XI, 499 pages. 1995.

Vol. 1013: T.W. Ling, A.O. Mendelzon, L. Vieille (Eds.), Deductive and Object-Oriented Databases. Proceedings, 1995. XIV, 557 pages. 1995.

Vol. 1014: A.P. del Pobil, M.A. Serna, Spatial Representation and Motion Planning. XII, 242 pages. 1995.

Vol. 1015: B. Blumenthal, J. Gornostaev, C. Unger (Eds.), Human-Computer Interaction. Proceedings, 1995. VIII, 203 pages. 1995.

Vol. 1017: M. Nagl (Ed.), Graph-Theoretic Concepts in Computer Science. Proceedings, 1995. XI, 406 pages. 1995.